The Power of Algorithms

Giorgio Ausiello • Rossella Petreschi
Editors

The Power of Algorithms

Inspiration and Examples in Everyday Life

Editors
Giorgio Ausiello
Dip. di Informatica e Sistemistica
Università di Roma "La Sapienza"
Rome, Italy

Rossella Petreschi
Dipartimento di Informatica
Università di Roma "La Sapienza"
Rome, Italy

First published in Italian in 2010 by Mondadori Education S.p.A., Milano as "L'informatica invisibile: Come gli algoritmi regolano la nostra vita ... e tutto il resto"

ISBN 978-3-642-39651-9   ISBN 978-3-642-39652-6 (eBook)
DOI 10.1007/978-3-642-39652-6
Springer Heidelberg New York Dordrecht London

Library of Congress Control Number: 2013952981

© Springer-Verlag Berlin Heidelberg 2013
This work is subject to copyright. All rights are reserved by the Publisher, whether the whole or part of the material is concerned, specifically the rights of translation, reprinting, reuse of illustrations, recitation, broadcasting, reproduction on microfilms or in any other physical way, and transmission or information storage and retrieval, electronic adaptation, computer software, or by similar or dissimilar methodology now known or hereafter developed. Exempted from this legal reservation are brief excerpts in connection with reviews or scholarly analysis or material supplied specifically for the purpose of being entered and executed on a computer system, for exclusive use by the purchaser of the work. Duplication of this publication or parts thereof is permitted only under the provisions of the Copyright Law of the Publisher's location, in its current version, and permission for use must always be obtained from Springer. Permissions for use may be obtained through RightsLink at the Copyright Clearance Center. Violations are liable to prosecution under the respective Copyright Law.
The use of general descriptive names, registered names, trademarks, service marks, etc. in this publication does not imply, even in the absence of a specific statement, that such names are exempt from the relevant protective laws and regulations and therefore free for general use.
While the advice and information in this book are believed to be true and accurate at the date of publication, neither the authors nor the editors nor the publisher can accept any legal responsibility for any errors or omissions that may be made. The publisher makes no warranty, express or implied, with respect to the material contained herein.

Printed on acid-free paper

Springer is part of Springer Science+Business Media (www.springer.com)

Preface

The meaning of the word *algorithm* as found in any English dictionary is rather similar to the meaning of words such as *method* or *procedure*, that is, "a finite set of rules specifying a sequence of operations to solve a particular problem". Simple algorithms we are all familiar with are those used to perform the four arithmetical operations, or the binary search which, more or less unconsciously, we use to find a name in a telephone directory.

Strangely, however, the very mention of the word algorithm provokes a sense of fear in many people, possibly due to its mathematical connotations. Indeed, the word's etymological origin is the name of the Persian mathematician, al-Khwarizmi, who worked in Baghdad at the beginning of the ninth century, and its contemporary meaning is derived from the fact that he introduced Indian methods of calculation based on positional representation of numbers into the Christian countries of the West.

And so it may be that a deep-seated unease with mathematics causes many to lose sight of the central role algorithms play in computer science and of the fact that myriad activities of their lives are today governed by algorithms. Booking a plane ticket, effecting a secure transaction at the cash machine of a bank, searching for information on the Web, and zipping or unzipping files containing music or images are just a few examples of the way algorithms have come to pervade all aspects of everyday life. Algorithms are even inserted into national legislation, such as the rules defining the construction of a citizen's fiscal code, national insurance number, etc., or the increasingly widely used digital signature for authenticating documents.

A highly important consideration to emphasize, however, is that not only do algorithms have a huge number of applications, but they also act as powerful "magnifying lenses" enabling a penetrating comprehension of problems.

Examining, analyzing, and manipulating a problem to the point of being able to design an algorithm leading to its solution is a mental exercise that can be of fundamental help in understanding a wide range of subjects, irrespective of the fields of knowledge to which they belong (natural sciences, linguistics, music, etc.).

In any case, it was the advent of computers and computer science that led to the word 'algorithm' becoming known to a wide range of people, so much so that even in 1977 Donald Knuth (one of the founding fathers of computer science) wrote:

> Until ten years ago the word algorithm was unknown to the vast majority of educated people and, to tell the truth, there was little need for it anyway. The furiously rapid development of computer science, whose primary focus is the study of algorithms, has changed this situation: today the word algorithm is indispensable.

Formalizing a problem as an algorithm thus leads to a better grasp of the argument to be dealt with, compared to tackling it using traditional reasoning. Indeed, a person who knows how to handle algorithms acquires a capacity for introspection that she/he will find useful not only in writing good programs for a computer, but also in achieving improved understanding of many other kinds of problem in other fields. Knuth, again, in his book "The Art of Computer Programming", asserts that:

> If it is true that one doesn't truly understand a problem in depth until one has to teach it to someone else, it is even truer that nothing is understood more completely than something one has to teach to a machine, that is, than something which has to be expressed by way of an algorithm.

Unfortunately, the precision demanded by the algorithmic approach (the algorithm has to be independent of the data to which it is applied and the rules it employs have to be elementary, that is, very simple and unambiguous), although useful as a means of mental development, limits the types of problem for which it can be adopted. To convince oneself of this just think of the fact that no algorithm exists for teaching "how to live a happy life". Alternatively, as a more rigorous demonstration of these limitations, we cite one of the most important findings of twentieth century logic, whereby Alan Turing (in the wake of Gödel's incompleteness proof) showed that no algorithm exists that would be capable of deciding whether or not a logical formula asserting a property of arithmetic is a theorem (see Chaps. 1 and 3).

For every algorithm two fundamental components can be identified: the determination of the appropriate algorithmic design technique (based on the structure of the problem) and the clear understanding of the mathematical nucleus of the problem. These two components interact closely with each other, thus it is not so much that algorithmic ideas just find solutions to well-stated problems, as that they function as a language that enables a particular problem to be expressed in the first place. It is for this reason that David Harel, in his 1987 book "Algorithmics: The Spirit of Computing" was able, without fear of contradiction, to define the algorithm as "the soul of computer science".

The earliest algorithms can be traced back as far as 2000 BCE; Mesopotamian clay tablets and Egyptian papyrus have been found bearing the first examples of procedures for calculation defined in fairly rigorous ways. Over the successive millennia thereafter humans made ever-increasing use of algorithms to solve problems arising in widely diverse fields: from measurements of land areas to astronomy, from trade to finance, and from the design of civil engineering projects to the study of physical phenomena. All of these significantly contributed, in the eighteenth and nineteenth centuries, to the first products of the industrial revolution.

Notwithstanding this, it was not until the twentieth century that the formal definition of the concept of algorithm began to be tackled. This was done primarily by mathematical logicians, such as Alonzo Church and the already-cited Alan Turing, in a series of theoretical investigations which turned out to be the indispensable groundwork for subsequent development of the first programmable electronic computer and the first computer programming languages. As mentioned earlier, it was with the advent of computers and computer science that algorithms really began to play a central role, initially only in military and scientific fields, and then ever increasingly in the fields of commerce and management. Today we can say that algorithms are an indispensable part of our everyday lives—and it seems they are destined to become even more pervasive in the future.

Nevertheless, despite this massive influence of algorithms on the world around us, the majority of users remain totally ignorant of their role and importance in securing the performance of the computer applications with which they are most familiar, or, at best, consider them technical matters of little concern to them. Instead quite the opposite is the case: in reality it is the power, the precision, the reliability and the speed of execution which these same users have been demanding with ever-increasing pressure that have transformed the design and construction of algorithms from a highly skilled "niche craft" into a full-fledged science in its own right.

This book is aimed at all those who, perhaps without realizing it, exploit the results of this new science, and it seeks to give them the opportunity to see what otherwise would remain hidden. There are ten chapters, of which nine are divided into two parts. Part I (Chaps. 1–3) introduces the reader to the properties and techniques upon which the design of an efficient algorithm is based and shows how the intrinsic complexity of a problem is tackled. Part II (Chaps. 4–9) presents six different applications (one for each chapter) which we encounter daily in our work or leisure routines. For each of these applications the conceptual and scientific bases upon which the algorithm used is grounded are revealed and it is shown how these bases are decisive as regards the validity of the applications dealt with. The book concludes with a different format, that of the dialogue. Chapter 10 illustrates how randomness can be exploited in order to solve complex problems, and its dialogue format has been deliberately chosen to show how discussions of such issues are part of the daily life of those who work in this field.

As an aid to readers whose educational background may not include particularly advanced mathematics there are clear indications in the text as to which sections containing more demanding mathematics may be skipped without fear of losing the thread of the main argument. Moreover, in almost every chapter, boxes covering specific mathematical or technical concepts have been inserted, and those readers wishing to get a general sense of the topic can avoid tackling these, at least on a first reading.

In fact, an overriding aim of the authors is to make the role of algorithms in today's world readily comprehensible to as wide a sector of the public as possible. To this end a simple, intuitive approach that keeps technical concepts to a minimum has been used throughout. This should ensure ideas are accessible to the intellectually curious reader whose general education is of a good level, but does not necessarily include mathematical and/or computer scientific training.

At the same time, the variety of subjects dealt with should make the book interesting to those who are familiar with computer technologies and applications, but who wish to deepen their knowledge of the ideas and techniques that underlie the creation and development of efficient algorithms. It is for these reasons that the book, while having a logical progression from the first page to the last, has been written in such a way that each chapter can be read separately from the others.

Roma, Italy
July 2013

Giorgio Ausiello
Rossella Petreschi

Contents

Part I Finding One's Way in a World of Algorithms

1 Algorithms, An Historical Perspective 3
Giorgio Ausiello
  1.1 Introduction 3
  1.2 Teaching Algorithms in Ancient Babylonia and Egypt 4
  1.3 Euclid's Algorithm 8
  1.4 Al-Khwarizmi and the Origin of the Word Algorithm 10
  1.5 Leonardo Fibonacci and Commercial Computing 13
  1.6 Recreational Algorithms: Between Magic and Games 17
  1.7 Algorithms, Reasoning and Computers 21
  1.8 Conclusion 25
  1.9 Bibliographic Notes 26

2 How to Design an Algorithm 27
Rossella Petreschi
  2.1 Introduction 27
  2.2 Graphs 28
    2.2.1 The Pervasiveness of Graphs 28
    2.2.2 The Origin of Graph Theory 32
    2.2.3 The Topological Ordering Problem 35
  2.3 Algorithmic Techniques 36
    2.3.1 The Backtrack Technique 37
    2.3.2 The Greedy Technique 42
  2.4 How to Measure the Goodness of an Algorithm 49
  2.5 The Design 52
  2.6 Bibliographic Notes 57

3 The One Million Dollars Problem 59
Alessandro Panconesi
  3.1 Paris, August 8, 1900 61
  3.2 *"Calculemus!"* 65

	3.3	Finding Is Hard: Checking Is Easy	67
	3.4	The Class NP	70
	3.5	Universality	74
	3.6	The Class P	74
	3.7	A Surprising Letter	76
	3.8	The Driving Force of Scientific Discovery	80
	3.9	Bibliographic Notes	80

Part II The Difficult Simplicity of Daily Life

4 The Quest for the Shortest Route 85
Camil Demetrescu and Giuseppe F. Italiano

4.1	Introduction	85
4.2	The Mathematisch Centrum	88
4.3	Shortest Paths in Graphs	89
4.4	Nature and Its Algorithms	90
4.5	A Simple Idea	91
4.6	Time Is a Tyrant	94
4.7	How to Set Your Priorities	96
	4.7.1 The Heap Data Structure	98
4.8	The Humble Programmer	100
4.9	Still an Open Challenge	101
	4.9.1 The ALT Algorithm by Goldberg and Harrelson	103
4.10	Bibliographic Notes	105

5 Web Search 107
Paolo Ferragina and Rossano Venturini

5.1	The Prologue	107
5.2	Internet and Web Graphs	108
5.3	Browsers and a Difficult Problem	114
5.4	Search Engines	118
	5.4.1 Crawling	120
	5.4.2 The Web Graph in More Detail	122
	5.4.3 Indexing and Searching	124
	5.4.4 Evaluating the Relevance of a Page	127
	5.4.5 Two Ranking Algorithms: PageRank and HITS	129
	5.4.6 On Other Search Engine Functionalities	133
5.5	Towards Semantic Searches	134
5.6	Bibliographic Notes	137

6 Algorithms for Secure Communication 139
Alberto Marchetti-Spaccamela

6.1	Introduction	139
6.2	A Brief History of Cryptography	141
	6.2.1 Monoalphabetic Substitution Codes	141

		6.2.2 Polyalphabetic Substitution Codes	143
		6.2.3 The Enigma Machine	144
	6.3	Cryptographic Codes and Secret Keys	145
		6.3.1 How to Encode a Long Message Using an Integer Function	146
		6.3.2 Cryptanalysis and Robustness of a Cryptographic Protocol	147
	6.4	Secret Key Cryptography	151
		6.4.1 Secret Key Cryptography Standards	151
		6.4.2 Limitations of Secret Key Encryption	152
	6.5	The Key Distribution Problem	153
		6.5.1 Modular Arithmetic	154
		6.5.2 Diffie and Hellman's Algorithm for Establishing a Secret Key	155
	6.6	Public-Key Cryptography	157
		6.6.1 The RSA Algorithm	158
	6.7	Digital Signatures and Other Useful Applications of Public-Key Cryptography	161
		6.7.1 How Public-Key Cryptography Allows for Digital Signatures	162
	6.8	Bibliographic Notes	165

7 Algorithmics for the Life Sciences ... 167
Raffaele Giancarlo

	7.1	Introduction	167
	7.2	The Fundamental Machinery of Living Organisms	170
	7.3	Algorithmic Paradigms: Methodological Contributions to the Development of Biology as an Information Science	174
		7.3.1 String Algorithmics: Identification of Transcription Factors Binding Sites	175
		7.3.2 Kolmogorov Algorithmic Complexity: Classification of Biological Sequences and Structures	178
		7.3.3 Graph Algorithmics I: Microarrays and Gene Expression Analysis	179
		7.3.4 Graph Algorithmics II: From Single Components Towards System Biology	182
	7.4	Future Challenges: The Fundamental Laws of Biology as an Information Science	184
	7.5	Bibliographic Notes	185

8 The Shortest Walk to Watch TV ... 187
Fabrizio Rossi, Antonio Sassano, and Stefano Smriglio

	8.1	A Different Idea of Television	187
	8.2	Designing a Broadcasting Network	189
		8.2.1 The Physical Elements of the Network	189
		8.2.2 Computer Representation	190

		8.2.3	Model for the Digital Coverage Assessment	191
		8.2.4	Network Design	194
	8.3	The Role of Transmission Delays		194
	8.4	An Algorithm for Optimizing Transmission Delays		199
		8.4.1	From Inconsistent TP Sets to Inconsistent Systems of Inequalities	200
		8.4.2	The Difference Constraints Graph	202
		8.4.3	Shortest Walks in G and Transmission Delays	203
	8.5	From Shortest Walk to Television		205
	8.6	Bibliographic Notes		205
9	**Algorithms for Auctions and Games**			**207**
	Vincenzo Bonifaci and Stefano Leonardi			
	9.1	Introduction		207
	9.2	Games and Solution Concepts		209
		9.2.1	Prisoner's Dilemma	209
		9.2.2	Coordination Games	211
		9.2.3	Randomized Strategies	212
		9.2.4	Hawks and Doves	214
	9.3	Computational Aspects of Game Theory		216
		9.3.1	Zero-Sum Games and Linear Optimization	216
		9.3.2	Fixed-Points: Nash's Theorem and Sperner's Lemma	218
		9.3.3	Mixed Nash Equilibria in Non-zero-Sum Games	221
	9.4	Inefficiencies		222
		9.4.1	The Tragedy of the Commons	222
		9.4.2	Routing Games	224
	9.5	Mechanism Design and Online Auctions		226
		9.5.1	The Vickrey Auction	226
		9.5.2	Vickrey–Clarke–Groves Mechanisms	228
		9.5.3	Computational Aspects of Mechanism Design	230
	9.6	Price-Setting Mechanisms and Competitive Equilibria		233
	9.7	Bibliographic Notes		234
10	**Randomness and Complexity**			**235**
	Riccardo Silvestri			
	10.1	A Dialogue		235
	10.2	Bibliographic Notes		250
References				**251**

List of Contributors

Giorgio Ausiello Dipartimento di Ingegneria Informatica, Automatica e Gestionale, Sapienza Università di Roma, Roma, Italy

Vincenzo Bonifaci Istituto di Analisi dei Sistemi ed Informatica "Antonio Ruberti", Consiglio Nazionale delle Ricerche, Roma, Italy

Camil Demetrescu Dipartimento di Ingegneria Informatica, Automatica e Gestionale, Sapienza Università di Roma, Roma, Italy

Paolo Ferragina Dipartimento di Informatica, Università di Pisa, Pisa, Italy

Raffaele Giancarlo Dipartimento di Matematica ed Informatica, Università di Palermo, Palermo, Italy

Giuseppe F. Italiano Dipartimento di Ingegneria Civile e Ingegneria Informatica, Università di Roma "Tor Vergata", Roma, Italy

Stefano Leonardi Dipartimento di Ingegneria Informatica, Automatica e Gestionale, Sapienza Università di Roma, Roma, Italy

Alberto Marchetti-Spaccamela Dipartimento di Ingegneria Informatica, Automatica e Gestionale, Sapienza Università di Roma, Roma, Italy

Alessandro Panconesi Dipartimento di Informatica, Sapienza Università di Roma, Roma, Italy

Rossella Petreschi Dipartimento di Informatica, Sapienza Università di Roma, Roma, Italy

Fabrizio Rossi Dipartimento di Informatica, Università dell'Aquila, Coppito (AQ), Italy

Antonio Sassano Dipartimento di Ingegneria Informatica, Automatica e Gestionale, Sapienza Università di Roma, Roma, Italy

Riccardo Silvestri Dipartimento di Informatica, Sapienza Università di Roma, Roma, Italy

Stefano Smriglio Dipartimento di Informatica, Università dell'Aquila, Coppito (AQ), Italy

Rossano Venturini Dipartimento di Informatica, Università di Pisa, Pisa, Italy

Part I
Finding One's Way in a World of Algorithms

Chapter 1
Algorithms, An Historical Perspective

Giorgio Ausiello

Abstract The design of algorithms for land measurement, financial transactions and astronomic computations goes back to the third millennium BCE. First examples of algorithms can be found in Mesopotamian tablets and in Egyptians scrolls. An important role in the development of numerical algorithms was played in the ninth century by the Persian mathematician al-Khwarizmi, who introduced the Indian numeration systems to the Arab world and from whom we derived the name 'algorithm' to denote computing procedures. In the Middle Ages algorithms for commercial transactions were widely used, but it was not until the nineteenth century that the problem of characterizing the power of algorithms was addressed. The precise definition of 'algorithm' and of the notion of computability were established by A.M. Turing in the 1930s. His work is also considered the beginning of the history of Computer Science.

1.1 Introduction

The ability to define algorithms for numerical computations or, more generally, as we would say today, for data processing, starts to appear in the history of mankind a few millennia before Christ. Among the most ancient examples of this ability are some tools used for taking note of the results of computations and, especially, the first calendars designed in ancient Egypt. In this chapter, far from attempting a history of algorithms, an effort that would require several volumes by itself, we want to show meaningful examples of algorithms, both numerical and non-numerical, that have been designed, studied and used throughout various historical ages. In the choice and illustration of such examples, there are two most relevant aspects that

---

G. Ausiello (✉)
Dipartimento di Ingegneria Informatica, Automatica e Gestionale, Sapienza Università di Roma, via Ariosto 25, 00185 Roma, Italy
e-mail: ausiello@dis.uniroma1.it

should be taken into account and that are still important nowadays in the design of modern algorithms. The first one derives from the very notion of algorithm, and corresponds to the need to find the correct sequence of precise and elementary operations that duly executed allow one to reach the solution of a problem in a finite number of steps. The second is related to the need to communicate to other people the sequence of computing steps to be performed and is related, therefore, to the use of a formal and unambiguous language in the presentation of an algorithm.

It is interesting to observe that these two properties (finiteness and formal definability) were understood only recently (less than a century ago, see Sect. 1.7) and are exactly the two properties that nowadays allow us to write, in a suitable programming language, algorithms that can be interpreted and performed by an electronic computing device (see Sect. 2.5). The same properties allow us to highlight the difference between the history of computing and the history of mathematics. It is clear, in fact, that in some sense the history of algorithms is part of the history of mathematics: various fields of mathematics developed due to the need to find solution methods for precise problems.[1] On the other hand are exactly the above-cited finiteness and constructiveness characters that draw a borderline with respect to those fields of mathematics (set theory, function theory, topology, etc.) in which, instead, the study and demonstration of properties of abstract structures have to employ the concepts of the infinitely small and the infinitely large, and often require the use of nonconstructive existential proofs.

1.2 Teaching Algorithms in Ancient Babylonia and Egypt

The oldest nontrivial example of numerical computation that we are aware of is reported on a Sumerian clay tablet from around 2500 BCE, found in Shuruppak, on the Euphrates river. In this example a simple basic problem is addressed, typically related to the life of an agricultural community: the subdivision of the content of a wheat warehouse among various persons in such a way that each person receives a specified amount of wheat. Hence the problem consists in computing how many people can receive their portion of wheat. Actually, the scribe does not present a particular algorithm but just the obtained result. Such a document is nevertheless interesting because we can derive from it information about the number system used by the Sumerians (a mixed decimal and sexagesimal system), and we learn that Sumerians knew various ways to execute division.

More interesting to understanding how algorithms were defined and used in ancient times are some Babylonian tablets from the period 2000 to 1650 BCE. In this case (as in the case of contemporary Egyptian scrolls), the algorithms are presented in a didascalic and repetitive style with reference to specific numerical examples.

---

[1]For example, Herodotus claimed that the development of geometry in ancient Egypt was due to the need to solve land measurement problems arising from repeated Nile floods.

1 Algorithms, An Historical Perspective

> The number is 4; 10. What is its inverse?
> Proceed as follows.
> Compute the inverse of 10. You will find 6.
> Multiply 6 by 4. You will find 24.
> Add 1. You will find 25.
> Compute the inverse of 25. You will find 2; 24.
> Multiply 2; 24 by 6. You will find 14; 24.
> The inverse is 14; 24. This is the way to proceed.

Fig. 1.1 Algorithm for the inversion of the number 4; 10 (that is 250). It is easy to check that 14; 24 is the inverse of 4; 10 since $(4 \times 60 + 10) \times (14 \times 60^{-2} + 24 \times 60^{-3}) = 1$

> The number is x. What is its inverse?
> Proceed as follows.
> [Let y and z be two numbers such that $x = y + z$]
> Compute the inverse of y. You will find y'.
> Multiply y' by z. You will find t.
> Add 1. You will find u.
> Compute the inverse of u. You will find u'.
> Multiply u' by y'. You will find v.
> The inverse is v. This is the way to proceed.

Fig. 1.2 Algorithm for the inversion of the number x. The algorithm is derived from the example in Fig. 1.1 by replacing the numbers appearing in the example with variables

What we can argue is that such tablets were used to teach algorithms to students, that is, to present them with the lists of elementary operations to be performed for each specific numerical example. The general rule (although not explicitly presented) could then be inductively derived from the examples. In the tablets various problems are addressed: square root computations, resolution of second-degree equations, computation of the inverse of a given number, etc. Let's look at one of the examples in detail. This will also offer us the possibility to observe the Babylonian number system more closely.

The problem consists in inverting a given number and, as we said above, the algorithm (Fig. 1.1) does not refer to a generic number denoted by a variable x as we would do today (and as we actually do in Fig. 1.2 in order to illustrate the computation executed by the algorithm), but to a specific value. Let us remember that the inverse of a number x is the number $y = 1/x$ that multiplied by x gives as result the number 1. For example, the inverse of 2 is 30 since choosing a suitable power of the basis 60 we have $2 \times 30 \times 60^{-1} = 1$. Computing the inverse is an important basic operation and, in particular, it was an important operation at that time, since the division operation was performed by multiplying the dividend by the inverse of the divisor. For simple numbers tables of inverses were available. In order to explain the algorithm we have to specify that, in the original text, numbers are represented in cuneiform characters and are expressed in mixed decimal and sexagesimal base. Every number consists of a sequence of values between 1 and 59, each one expressed in base 10 that here, for the sake of clarity, we present separated by the symbol ";". The number zero simply corresponds to an empty space. The sequence 2; 4; 10, for example, denotes the number $2 \times 60^2 + 4 \times 60 + 10 = 7,200 + 240 + 10 = 7,450$. Which powers of 60 were used depended on the context, therefore the same sequence might represent the number $2 \times 60^3 + 4 \times 60^2 + 10 \times 60$. In particular, and this has to be regarded as one of the advanced characteristics of Babylonian mathematics, the same notation might be used to express decimal numbers. For example, the above-mentioned sequence 2; 4; 10 might as well represent the number $2 + 4 \times 60^{-1} + 10 \times 60^{-2}$.

In Fig. 1.1 we can see how the hypothetical Babylonian teacher could present the algorithm for the computation of the inverse of the number 4; 10 (corresponding to 250) to his students. The presentation starts as follows: "The number is 4; 10. What is its inverse?" In the same tablet the algorithm is presented several times, each time applied to different numbers expressed in sexagesimal notation. For example, "The number is 8; 20. What is its inverse?", "The number is 1; 13; 20. What is its inverse?", etc.

It is interesting to pay attention to the last sentence in the presentation of the algorithm: "This is the way to proceed." This sentence shows that the person writing the text was conscious of having discovered a computation procedure, in other words an algorithm, to solve the general problem. In fact, the procedure always followed the same steps, independently from the input values. By means of a simple abstraction process it is easy for us to derive the underlying algorithm from the examples (see Fig. 1.2). The computation method is based on the expression $1/x = 1/(y+z) = 1/y \times 1/(z \times 1/y + 1)$ and consists in reducing the computation of the inverse of a given number x to the computation of the inverse of the two smaller numbers y and $u = z \times 1/y + 1$ until we reach numbers for which the inverse is already known (or can be found in precomputed tables of inverses[2]).

A very similar approach in the presentation of algorithms can be found in a famous Egyptian papyrus belonging to the first centuries of the second millennium BCE. The scroll is known as the "Rhind papyrus" from the name of a Scottish traveler who bought some of its fragments, or the "Ahmes papyrus" from the name of the scribe who copied it from an older document. This papyrus is currently held in the British Museum (with the name pBM 10057) and, together with the so-called "Moscow papyrus" and a leather scroll also held in the British Museum, is one of the few documents that provide us with information about the mathematical knowledge in ancient Egypt. Despite its ambitious title, "Accurate reckoning for inquiring into the nature and into the knowledge of all things, all mysteries, all secrets", the document just contains a collection of examples showing how the computation should be carried out in particular cases. Again, as in the case of the Babylonian tablets, the general computation rules (the algorithms) are not explicitly provided in the document, but we can easily infer them from the examples as we did above for the computation of the inverse. The examples provided in the papyrus concern

---

[2]Babylonians left several tables of simple inverses.

1 Algorithms, An Historical Perspective

> Example of the computation of a triangle of land surface.
> If you are told: a triangle is high 10 khet and his base is 4 khet.
> What is its area? Do as it has to be done.
> Divide 4 by 2. You obtain 2.
> Multiply 10 by 2. This is its area.
> Its area is 20.

Fig. 1.3 Problem 51 in the Rhind papyrus: algorithm for the computation of the area of a triangle of height 10 and base 4

$$
\begin{array}{ll}
34 \times 1 = 34 & 21 : 2 = 10 \text{ remainder } 1 \\
34 \times 2 = 34 + 34 = 68 & 10 : 2 = 5 \text{ remainder } 0 \\
34 \times 4 = 68 + 68 = 136 & 5 : 2 = 2 \text{ remainder } 1 \\
34 \times 8 = 136 + 136 = 272 & 2 : 2 = 1 \text{ remainder } 0 \\
34 \times 16 = 272 + 272 = 544 & 1 : 2 = 0 \text{ remainder } 1 \\
\end{array}
$$
$$34 \times 21 = 34 \times (1 + 4 + 16) = 34 + 136 + 544 = 714$$

Fig. 1.4 Multiplication of 34 by 21 with the method of duplicating the first factor and halving the second factor. Note that the sequence 10101 is the binary representation of 21

a large variety of problems: computation of fractions, computation of geometrical series, resolution of simple algebraic equations, and computation of surfaces and volumes.

In order to illustrate the way algorithms are presented in the Rhind papyrus, let us choose a simple example for the computation of the area of a triangle. The sequence of computation steps (somewhat rephrased) is presented in Fig. 1.3.

It is interesting to observe that the style of presentation of the algorithm is quite similar to the style we have found in Babylonian tablets. It is also worth noting that, again in this case, the author is aware of the paradigmatic character of the computation procedure presented in the example when he says: "Do as it has to be done".

Among the algorithms presented in the Rhind papyrus it is particularly relevant to cite the multiplication algorithm based on the so-called technique "by duplicating and halving". The technique is based on the distributive property of multiplication and on the possibility to represent a number as the sum of powers of two (the same property that is behind the binary representation of numbers in computers). Let us consider the following example: Suppose we have to multiply 34 by 21. The same result can be obtained by computing $34 \times (1 + 4 + 16)$ with the advantage that the product of 34 by a power of two can be obtained by means of repeated sums (duplications: see Fig. 1.4). As a consequence, although less efficient than the multiplication algorithm that we use nowadays, the described method does not require knowing the multiplication table (the so-called "table of Pythagoras").

1.3 Euclid's Algorithm

In various beginners' classes in mathematics and computer science, one of the first algorithms that is taught is also one of the most ancient: Euclid's algorithm for the computation of the greatest common divisor of two integer numbers.

The algorithm is presented in book VII of the Elements, Euclid's main work. For over two millennia this book has been a fundamental source for mathematical studies, particularly for geometry and number theory. In the landscape of Greek mathematics, Euclid's algorithm plays a singular role. In fact, in contrast to the kind of mathematics used by Egyptians and Mesopotamian peoples, oriented, as we saw, to the solution of practical problems, Greek mathematics, starting with Thales' work, followed an abstract approach, based on a line of thought that nowadays we would call axiomatic and deductive. On one side, this approach was a big cultural leap and influenced the future course of the discipline, but, on the other side, it put in a secondary place the aspects related to computation and algorithm design. The algorithm for greatest common divisor computation, therefore, is in a sense an exception. At the same time, it has to be noted that the style in which the algorithm is presented by Euclid offers an important step forward with respect to the way in which the computation processes we have seen until now were presented. In Euclid's text, in fact, the algorithm is formulated in abstract terms, with reference to arbitrary values and is not applied to specific integer numbers given as examples. In addition, the algorithm is formulated in geometrical terms. The arbitrary integer values are actually represented by means of segments and the expression "a number measures another number" expresses the idea that the smaller number divides the larger number and can, therefore, be adopted as a unit of measure of the larger number (just as a shorter stick can be used to measure the length of a longer stick).

After defining the concept of relatively prime numbers (segments that have the property that the only segment that divides them both is the unit segment), Euclid proceeds to defining the computation process in the following terms.

> Suppose we are given two numbers AB and $\Gamma\Delta$ that are not relatively prime, and suppose that $\Gamma\Delta$ is the smaller. We have to find the greatest common measure of the numbers AB and $\Gamma\Delta$. If $\Gamma\Delta$ measures AB then $\Gamma\Delta$ is the common measure of AB and $\Gamma\Delta$, since $\Gamma\Delta$ is also a measure of itself. Clearly it is also the greatest common measure since no number larger than $\Gamma\Delta$ can measure $\Gamma\Delta$. In case $\Gamma\Delta$ does not measure AB, if we eliminate $\Gamma\Delta$ from AB a number will exist that measures what is left. The remainder cannot be 1 since otherwise the numbers AB and $\Gamma\Delta$ would have been relatively prime, which is not in the hypothesis. Let us suppose that, measuring AB, $\Gamma\Delta$ leaves the remainder AE smaller than itself and let us also suppose that measuring $\Gamma\Delta$, AE leaves ΓZ smaller than itself. Let us finally suppose that ΓZ measures AE. Then, since ΓZ measures AE and AE measures $Z\Delta$, ΓZ measures $Z\Delta$. But since ΓZ is also a measure of itself, it follows that ΓZ measures the entire $\Gamma\Delta$. Now, since $\Gamma\Delta$ measures BE then ΓZ measures BE and since ΓZ also measures AE, this means that ΓZ measures the entire AB. Hence ΓZ is a common measure of both AB and $\Gamma\Delta$.

The text then goes on to prove that ΓZ is the largest common measure of AB and $\Gamma\Delta$ and finally ends with the statement *"This is what was to be proven"* (Fig. 1.5).

1 Algorithms, An Historical Perspective

```
A_____E_____B
Γ____Z_____Δ
```

Fig. 1.5 Computation of the greatest common divisor between the length of the segment AB and the length of the segment $\Gamma\Delta$. The result is given by the length of the segment ΓZ

Input: two integer numbers n and m.	
Output: the GCD of n and m.	
Step 1: If $m = n$ then	$GCD(n,m) = n$
Step 2: else	if $m > n$ then compute $GCD(m-n,n)$
	else compute $GCD(n-m,m)$

Fig. 1.6 Euclid's algorithm for the computation of the greatest common divisor (GCD)

In modern terms the algorithm can be more easily formulated in the way it appears in Fig. 1.6.

In the algorithm presented in Fig. 1.6 we use a technique that is widely employed in current programming languages and is now known as recursion. This technique (that as we saw was already implicitly used by Euclid) is inspired by the logical concept of induction (see Chap. 2) and consists in determining the value of a function applied to given arguments (in our case, the function GCD applied to n and m) by making use of the value that the same function would return when applied to smaller arguments ($m - n$ and n or, alternatively, $n - m$ and m).

It is also interesting to observe that Euclid's text provides, at the same time, both the algorithm to be followed for computing the function with arbitrary input values and the proof of its correctness (whose presentation is made easy thanks to the recursion approach used by the algorithm). This appears to be a great step forward, not only with respect to the way the algorithms we saw in the preceding sections were presented, but also with respect to the way algorithms are presented today, often without the support of a rigorous correctness proof. In order to guarantee the correct behavior of computer applications, in fact, it would be appropriate that both the algorithms used and the computer programs that implement them in applications were accompanied by formal proofs of their correctness (see Sect. 2.5). Unfortunately, this only happens rarely, and only for the most sophisticated applications.[3]

---

[3] In some cases, as users unfortunately realize at their own expense, for economy reasons, programs are written by poorly qualified personnel, without making use of the scientific programming methods that research in computer science has made available. This is why computer programs can sometimes behave differently than they were expected to and can even make errors with disastrous consequences.

1.4 Al-Khwarizmi and the Origin of the Word Algorithm

Few people realize that, when in 772 CE the seat of the caliphate was moved from Damascus to Baghdad, this had an exceptional impact on the history of mathematics and of algorithms. In fact, under the Abassid caliphs, and in particular during the caliphate of al-Mansur, Harun ar-Rashid (the legendary caliph of "One Thousand and One Nights"), al-Mamun and al-Mutasim, Baghdad became a very important center for the development of mathematics and science. The translation into Arabic of Greek scientific works had already started in the sixth and seventh centuries, but the early Islamic period also witnessed various violent actions carried out by fanatic religious people that even led to the destruction of books and other scientific works. During the Abassid caliphate, a more open-minded and rational point of view prevailed. Knowledge was gathered from all regions of the known world, processed and elaborated through a synthetic approach that allowed mathematicians and scientists to realize meaningful progress in all fields.

The last decades of the eighth century and the first decades of the ninth century were a flourishing period for Baghdad from both the economical and the cultural points of view. Not only did revenues from commercial exchanges flow to Baghdad from all regions reached by Arab merchants and ambassadors but also many manuscripts were collected in the many libraries of the city. Around the year 820, al-Mamun founded a scientific academy, the House of Wisdom (Bayt al-Hikma), which consisted of a library and an astronomy observatory where scientists and scholars in all disciplines were invited from abroad.

Among the mathematicians who arrived at the House of Wisdom was the person whose name was given to computational procedures: Abdallah Mohamed Ibn Musa al-Khwarizmi al-Magusi. Although we have limited information about his life, we know that he lived approximately between the years 780 and 850 and that, as his name reveals, he was the son of Musa and was born in Khoresme on the border between the regions that today belong to Iran and Uzbekistan. Recently the Uzbek government dedicated a stamp to him and a statue was placed in what is presumed to be his home town, Khiwa (Fig. 1.7).

It is interesting to observe that, as al-Khwarizmi himself reports, the task assigned to him by al-Mamun upon his arrival in Baghdad was mainly of a practical nature:

> ... to compose a short report concerning computation by means of the rules of restoration and reduction, limited to the simplest and most useful aspects of mathematics that are constantly applied to inheritances, legacies, their sharing out, court decisions and commercial transactions and in all other human business or when land measurements are required, excavations of water channels, geometric computations and similar things.

What al-Khwarizmi really did was of much greater impact: his works were of fundamental relevance in the development of arithmetics and over the centuries have been translated and disseminated, establishing the foundations of medieval mathematical thought.

Several works of al-Khwarizmi reached us through Medieval Latin translations: a treatise on mathematics, one on algebra, one on astronomy, one on geography,

Fig. 1.7 Uzbek stamp representing al-Khwarizmi

and a Hebrew calendar. We suspect that other works have been lost. In particular, al-Khwarizmi's name is related to the introduction of the positional decimal system in Islamic countries (and from there to Christian Europe). As is well known, in such systems, developed first in India, a fundamental role was played by the representation of zero by means of a special symbol, a small circle or a dot. Even if the decimal positional notation was in fact known for a long time (as we have also seen, the Mesopotamians used in a sense a positional system), the an arithmetic treatise by al-Khwarizmi was the first mathematical work to provide a detailed presentation of the rules for executing the four basic operations and for computing with fractions according to such notation. The efficiency of such computing methods, compared to the less efficient methods based on the use of the abacus, determined the dissemination of the Indian numbering system (that indeed we call the "Arabic numbering system") and contributed to making the name of al-Khwarizmi famous and to the use of his name to denote any kind of algorithm.

In addition, al-Khwarizmi's treatise devoted to algebra (*Book of algebra and al-muqabala*) had an important role in the development of mathematical knowledge. In this treatise several algebraic problems are presented, most of which derived from applications (e.g., subdivision of legacies), in particular, a series of algorithms for the solution of first- and second-degree equations with numerical coefficients, duly organized into six different classes according to their structure.

In Fig. 1.8 we provide a simple example of the method used to solve the equation $x^2 + 10x = 39$ (this example would later appear in several medieval algebra textbooks). The method is called "square completion" and, in the particular case of our example, consists in constructing first a square of size x (the unknown value)

Fig. 1.8 Geometric method for solving the equation $x^2 + 10x = 39$

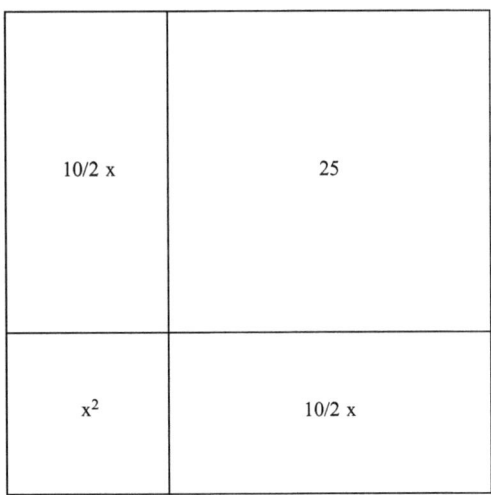

on whose sides two rectangles of sides x and $10/2$ are built. Finally the larger square is completed by introducing the square with side $10/2$. The larger square that we have constructed in this way has sides of size $x + 5$ and area of size $x^2 + 10x + 25$. But according to what is established by our equation we have: $x^2 + 10x + 25 = 39 + 25 = 64$, and therefore the sides of the larger square have size 8. From $x + 5 = 8$ we may derive $x = 3$.

As we said before, the works of al-Khwarizmi reached us through subsequent Latin versions, and the works of other mathematicians that were inspired by his texts. The number of Latin authors who spread the work of al-Khwarizmi is very large: John of Toledo (*Liber Algorismi de practice arismeticae*), Adelard of Bath (*Liber Ysagogarum Alchorismi in artem astronomicam a magistro A. compositus*), Leonardo Pisano (*Liber abbaci*), Alexander of Villadieu (*Carmen de Algorismo*), John of Halifax, better known as Sacrobosco, (*Algorismus vulgaris*), etc.

It is due to all these authors that the term *Algorismus* eventually became synonymous with computing procedures. For a long time the term was applied only with reference to arithmetical operations, as opposed to computing methods based on the use of abacus. In the *Florence Chronicle* written by Giovanni Villani, for example, we can read that in 1338, in Florence

> we find that boys and girls that learn to read are between eight and ten thousand. The young students that learn the abacus and the algorismus in six schools are between one thousand and one thousand two hundred.

Figure 1.9 shows that, according to an image contained in a book printed in 1508, the contraposition between algorithmic computations and computations based on the use of abacus was still present at the beginning of the sixteenth century.

Only in the eighteenth century did the term 'algorithm' start assuming the broad meaning that it has today. The Encyclopedia of d'Alambert and Diderot provides a definition of the term algorithm as follows:

1 Algorithms, An Historical Perspective 13

Fig. 1.9 The contraposition between algorithmic computations and computations based on the use of the abacus, as shown in [96]

Arab term used by some authors and in particular by Spanish authors to identify the practice of Algebra. Sometimes it is also applied to arithmetic operations based on digits. The same word is more generally used to denote method and notation of whatever kind of computation. In this sense we speak of algorithm for integral calculus, algorithm for exponential calculus, algorithm for sine calculus etc.

1.5 Leonardo Fibonacci and Commercial Computing

Among the mathematicians who contributed to spreading al-Khwarizmi's work, a very special role was played by Leonardo Pisano, also known as Fibonacci, who lived between c. 1180 and 1250, whose name is now, for various reasons, famous in the algorithm community (see Fig. 1.10). The main information concerning his origin and his life can be found in his most important work: the *Liber abbaci* that was written in 1202. Fibonacci was the son of a merchant from Pisa who worked in

Fig. 1.10 Leonardo Pisano, also known as Fibonacci

the important warehouse of Bugia (nowadays Béjaïa in Algeria), where he moved in 1192. In this merchant environment Fibonacci learned the algorithms of "Indian arithmetics" ("*Ubi ex mirabilis magisterio in arte per novem figuras Indorum introductus...*") that were extremely effective for commercial computing. Later, as a merchant, he traveled to various places around the Mediterranean (Egypt, Syria, Greece, Sicily and Provence), where he had the chance to upgrade his knowledge of mathematics. In the *Liber abbaci* he explains how he was able to integrate the knowledge of Indian and Arab computing (which he learned from al-Khwarizmi's works) with Euclidean mathematics. Fibonacci was also the author of other treatises such as the *Practica geometriae* and the *Liber quadratorum*. Although, from a mathematical point of view, the last one is probably his most original work, the *Liber abbaci* is undoubtedly the most relevant for its didactic value and for its role in making the name of Fibonacci famous throughout the Western world.[4]

Among the fifteen chapters of the *Liber abbaci* several concern problems of commercial nature; for example, the title of Chap. 9 is "De baractis rerum venalium", and the title of Chap. 10 is "De societatis factis inter consocios". In the volume various practical accounting problems (money exchange, computation of interest, amortization of debts, etc.) are addressed and make this work an important step in the history of accountancy.[5] In any case, the *Liber abbaci* cannot be classified

---

[4]The name of the volume should not be misunderstood: the book is entirely devoted to the Indo-Arabic computing system and the use of the abacus is never addressed.

[5]Thanks to his competence in this field, in 1241 Fibonacci was in charge of reorganizing the public accounting of the city of Pisa.

1 Algorithms, An Historical Perspective

Input:	integer n.	
Output:	$F(n)$: nth Fibonacci number.	
Step 1:	If $n = 0$ then	$F(n) = 1$
	Else if $n = 1$ then	$F(n) = 1$
	Else	$F(n) = F(n-1) + F(n-2)$

Fig. 1.11 Algorithm for the computation of Fibonacci numbers

just as an accountancy textbook nor as a handbook of commercial practice (as is the case of other books of the fourteenth and fifteenth centuries). This work is a real mathematical treatise that spans a great variety of topics, from integer arithmetic to fractional computing, from geometry to the solution of algebraic equations, from the computation of arithmetic and geometric series to the calculus of roots of equations.

The 12th chapter of the *Liber abbaci* ("De solutionibus multarum positarum questionum") is a rich source of mathematical problems, especially in the field of recreational mathematics. Among them we can find the famous "rabbit problem" that, as we will see, gained an important role in the history of algorithms: "Quot paria coniculorum in uno anno ex uno pario germinentur". In English the statement of the problem is more or less as follows: a man has a pair of rabbits in a secluded place and we would like to know how many rabbits this pair would generate in 1 year, taking into account that they can generate a pair every month and that after 1 month also the newly born can reproduce.

Fibonacci presents the solution in the following terms:

> Since in the second month the pair generates we will have two pairs in one month. One of these pairs (the first one) generates also in the second month, therefore we will have 3 pairs. After one more month two of them will be fertile and 2 pairs will therefore be born in the third month. We have then 5 pairs. Three of them will then be fertile and hence in the fourth month 3 more pairs will be born and 8 will be the overall number. The last month we will have 377 pairs. You can see in the margin how we operated: we summed the first number with the second then the second with the third, the third with the fourth... in this way you can compute for an infinite number of months.

It is easy to see that the computation procedure can be naturally formulated in recursive terms. The nth value of the sequence is defined as the sum of the $(n-1)$th and the $(n-2)$th values. Usually it is assumed that the value for $n = 0$ and $n = 1$ is 1, and hence the sequence is 1, 1, 2, 3, 5, 8, 13, 21, 34, 55, ... In modern terms the sequence (now known as the sequence of Fibonacci numbers) would be defined as in Fig. 1.11.

Beside being defined with the algorithm in Fig. 1.11, Fibonacci numbers can also be expressed in explicit terms by means of the expression: $F(n) = c_1((1 + \sqrt{5})/2)^n + c_2((1 - \sqrt{5})/2)^n$ with suitable coefficients c_1 and c_2 derived from the initial conditions $F(0)$ and $F(1)$.

Fibonacci numbers have several interesting properties. The most important one is that the ratio between $F(n)$ and $F(n-1)$ tends to the constant value $\phi = 1.618$, known as "the mean of Phidias" or the "golden ratio".

Fig. 1.12 Mario Merz, *The Flight of Numbers*, installation

It is worth remembering that the golden ratio is the ratio existing between a segment of unit length and a portion r of it that satisfies the relation $1 : r = r : 1-r$ (in other words, r is defined as the solution of the equation $r^2 + r = 1$). The value $r = 0.618$ (equal to $1/\phi$) is called the "golden section" of the segment.

The golden ratio was used by the ancient Greeks in order to obtain particularly elegant proportions between the dimensions of a building (such as, for example, a temple). More recently the existence of this harmonious ratio between a Fibonacci number and the number that precedes it in the sequence is at the base of the important role that Fibonacci numbers had in architecture for establishing the size of housing modules (Le Corbusier), in music for creating new tonal scales (Stockhausen), and in artistic installations (Merz, see Fig. 1.12).

This very same property is at the root of another reason why the name of Fibonacci is so important in computer science, beside, of course, the role that he played in disseminating Indo-Arabic computation methods. In fact, various data structures (see Fibonacci Heaps in Chap. 4) that allow us to perform operations on large data sets efficiently derive their name from Fibonacci, in particular, those data structures that allow us to handle a set of n records in logarithmic time.[6]

---

[6]In order to understand the great advantage to using this type of data structures it is enough to observe that searching and updating a database consisting of 500 million records can be done in fewer than 30 steps by making use of Fibonacci trees.

Fig. 1.13 Magic square of order 7

22	47	16	41	10	35	4
5	23	48	17	42	29	29
30	6	24	49	18	12	12
13	31	7	25	43	19	37
38	14	32	1	26	44	20
21	39	8	33	2	27	45
46	15	40	9	34	3	28

1.6 Recreational Algorithms: Between Magic and Games

As we saw in previous sections, throughout the centuries algorithms were developed mostly for practical needs, that is, for solving problems related to the productive activities of humans (agriculture, commerce, construction of buildings). The rabbit problem that is at the origin of the Fibonacci sequence of numbers and of the related algorithm is of a different nature. In this case we see an example that can be classified as recreational algorithmics, in which the design of algorithms is aimed at solving games and puzzles.

Other famous examples of algorithms of this nature are those referring to relational structures (which in the modern language of mathematics and computer science are called graphs), that since the eighteenth century have attracted the attention of mathematicians. As we will see in Chap. 2, such structures were initially defined and studied to solve recreational problems such as the classic problem introduced by Euler, which consisted of deciding whether it was possible to visit all parts of the city of Königsberg going exactly once over all seven bridges that connected them.

In this section we will address two other kinds of problems of a recreational nature that have been studied over the centuries and that stand at the intriguing crossroads between games, religion, and magic: the construction of magic squares and the traversal of labyrinths.

A magic square is a square table containing one integer in each cell arranged in such a way that the result obtained by summing up the integers contained in the cells of each row, of each column, and of each of the two diagonals, is, magically, the same. If the table contains n rows and n columns, the resulting magic square is said to be of "order n". Normally a magic square of order n contains all integers between 1 and n^2, and the magic sum is therefore equal to the sum of the first n^2 integers divided by n, that is $n(n^2 + 1)/2$ (see Induction in Chap. 2). In Fig. 1.13 a magic square of order 7 is shown, where the sum over each row, each column, and each of the two diagonals is equal to 175.

The simplest magic square is of order 3, since no magic square of order 2 can exist. Besides, it is in a sense unique because all the other eight magic squares of order 3 can be obtained by means of symmetries and rotations of the same square.

Fig. 1.14 The diagram of the Luo river (China, tenth century)

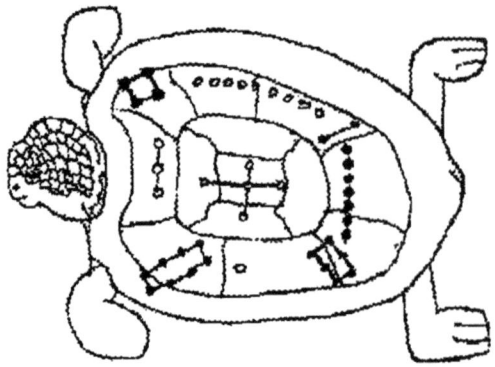

Such a square is also one of the oldest known magic squares; in fact, versions of this square have been known since the early Chinese Song dynasty (tenth century). The example in Fig. 1.14 is known as the Lo Shu magic square (literally the magic square of the Luo river) and is related to feng shui geomancy. There are 16 magic squares of order 4 and, from them, by means of symmetries and rotations, 880 variants can be obtained. The number of magic squares of order 5 is $275, 305, 224$. At present the exact number of magic squares of order 6 is not known.

As we said above, in various historical contexts, magic squares have taken a religious, esoteric, or even curative meaning, and for this reason they were called "magic", probably in the seventeenth century. Still, it is clear that the most interesting properties of such squares are the mathematical properties that since the eleventh century attracted the interest of Arab mathematicians like, for example, al-Haytham. In 1654 Pascal presented a work with the strange title "Treatise on magically magic numbers" to the Paris Academy and, subsequently, in the eighteenth and nineteenth centuries other famous mathematicians such as Fermat, Euler and Cayley analyzed the numerical and algebraic properties of magic squares.

It is not surprising that the interest in magic squares naturally led to the definition of algorithms for their construction. One of the most studied techniques is based on the progressive filling of concentric layers, moving from the outside toward the inside. This technique was formulated for the first time in the thirteenth century by the Arab mathematician al-Zinjani in his "Epistle on the numbers of harmony" and was developed in 1667 by the French scientist Arnauld. We cannot provide details on how the algorithm works here, but it is interesting to quote some comments that the author made on the algorithm. Such comments evidence the attention for the main properties that algorithms should satisfy, simplicity, generality, and efficiency, but also the possibility to demonstrate the correctness of the proposed technique.

> From this I think I can conclude that there does not exist a method which is easier, more synthetic and more perfect for constructing magic squares, one of the most beautiful arithmetical problems. What is peculiar with this method is that figures are written at most twice; we do not proceed by trials but we are always sure that what we do is correct; the largest squares are not more difficult to construct than smaller squares; various solutions can be obtained; nothing is done that cannot be demonstrated.

1 Algorithms, An Historical Perspective

o			o
	o	o	
	o	o	
o			o

4			1
	7	6	
	11	10	
16			13

4	14	15	1
9	7	6	12
5	11	10	8
16	2	3	13

Fig. 1.15 Construction of a magic square with the marking technique

A second algorithmic technique for constructing magic squares was proposed in the fourteenth century by the Byzantine mathematician Manuel Moschopoulos; it consists of filling a square of odd order n by progressively inserting all integers from 1 to n^2 following suitable rules, similar to the way in which chess pieces are moved on the chess board. Such rules can be expressed formally by making use of modular arithmetic. For example, one of the fundamental rules says that if integer x is in the cell corresponding to row i and column j, if x is not a multiple of n, the number $x + 1$ goes in row $i + 1$ (mod n)[7] and column $j + 1$ (mod n), while in the other case (when x is a multiple of n) $x + 1$ goes to row $i + 2$ (mod n) and column j.

The last technique that we want to mention is the one reported by Ibn Qunfudh, in the fourteenth century, in his work *The revelation of operations of computing*. Ibn Qunfudh's method is essentially a marking technique, which is likely to have been introduced centuries before and that can only be applied in the case of squares whose order is a multiple of 4. The method consists in marking, in a suitable way, one half of the cells of a square in such a way that exactly one half of the cells are marked on each row and on each column. Then, starting from the topmost cell on the right (and proceeding row by row from right to left) all integers from 1 to n^2 are progressively enumerated, and those corresponding to the marked cells are inserted in such cells. Finally, starting from the lowest cell on the left, in a similar way, all numbers from 1 to n^2 are again enumerated, and those corresponding to the empty cells are inserted in such cells (Fig. 1.15).

Another class of non-numeric algorithms that were frequently mentioned over the centuries and that have always fascinated experts in recreational mathematics are algorithms for the traversal of labyrinths.

As is well known, just as for magic squares, labyrinths also have taken important meanings and played diverse roles in ancient civilizations. Starting from the most ancient one, the labyrinth of el-Fayum, described by Herodotus, to the mythological labyrinth of the Minotaur, which the archeologist Arthur Evans identified with the Knossos royal palace in Crete, labyrinths have been mostly associated with the symbol of absolute power, the power that it is impossible to reach and, from which,

---

[7] We have to remember that the value a (mod b) corresponds to the remainder of a divided by b; for example 7 (mod 5) = 2.

Fig. 1.16 Labyrinth designed on the floor of Chartres cathedral

at the same time, it is impossible to escape. Other meanings have been assigned to labyrinths in religion (see the labyrinths designed on the floor of cathedrals to mean a trail of expiation, as in Fig. 1.16), in magic (labyrinths designed on jewels and amulets), in love (as a metaphor for the pains of the lover), and in architecture (labyrinths of hedges in Renaissance gardens).

We will not devote space here to the description of the various topological structures of labyrinths, nor will we address the issue of labyrinth design, which stimulated several studies across the centuries. We will instead concentrate on the issue of the traversal of labyrinths, either with the aim of reaching the center or of finding the way out.

As far as we know, the first real traversal algorithm was proposed by M. Trémaux in 1892. Technically it is an algorithm for "depth-first" visits, similar to the well-known one based on the idea of following the right wall by always maintaining the right hand on the wall. Different from that one, this algorithm can also be applied in the case where the labyrinth contains circular paths.

In the words of W.H. Matthews, who at the beginning of the twentieth century wrote a book on mazes and labyrinths, Trémaux's algorithm is as follows:

> On arriving at a node which, by the absence of marks, you know you have not already visited, mark the path by which you have just arrived by three marks; if you see by marks on other paths that you have already been to that node, mark the arrival path with one mark only. If now there are no unmarked paths at this node, it means that you have explored

this particular branch-system and must retrace your steps by the path by which you have arrived. If, however, there are one or more unmarked paths leading from the node, select one of them, and, as you enter it, mark it with two marks. On arrival at a node, you shall never take a path with three marks, unless there are no paths unmarked or with one mark only. When you enter a one-mark path, you shall add two marks so that it is now marked with three marks.

Speaking about Trémaux's algorithm it is nice to remember that the novelist Umberto Eco, in his well-known book *The Name of the Rose*, pretends that the algorithm was already known to the characters of his book in 1327 (the year in which the story is supposed to have happened) but, ironically, he tells us that when Adso of Melk asks *"Does this method allow us to escape from the labyrinth?"*, William of Baskerville replies *"Almost never, as far as I know."*

In conclusion, let us observe that the problem of traversing a labyrinth is still considered one of the fundamental problems in computer science. In fact, it has a paradigmatic value since finding the path between the entrance and the exit of a labyrinth is not very different from finding the connection between two entities in a relational structure, and this is a problem that frequently occurs in several computer applications.

1.7 Algorithms, Reasoning and Computers

Among the various applications of algorithms to human activities, the application to reasoning is certainly, for the reason we will examine in this section, the one that has the most relevant consequences. Automatic reasoning is still, definitely, one of the most ambitious and interesting research domains of artificial intelligence: together with the development of cognitive robotics, the applications of computers to reasoning will have a great impact on the future of mankind. What we want to emphasize here is that even the invention of the first computers derives, although indirectly, from the attempt to transform reasoning into an algorithmic process.

The possibility to realize algorithms that allow one to determine whether a sentence is true or false, and, in such a case, to derive from it all its logical consequences, started to appear in the Western world of philosophy in the seventeenth century, as a consequence of the success of mathematical computing and of the creation of the first computing machines (Pascal's machine is from 1642). Such an ambitious goal arose from the combination of two intellectual dreams that had been cultivated by philosophers from the time of Aristotle and that were reanimated in the Renaissance. The first one was the idea of recording and classifying all universal knowledge, by creating languages and ontologies able to represent all aspects of reality; the second consisted (so simply!) of the search for truth.

The German philosopher Gottfried Leibniz formulated the most ambitious proposal in this direction, based on his interdisciplinary culture, ranging from Aristotelian philosophy to combinatorics and from law to infinitesimal calculus, and animated by an optimistic (almost illuministic) spirit. He thought that it might be possible to organize all human knowledge in an "encyclopedia" and to apply to knowledge and reasoning the same computation rules that had been successfully applied to mathematics. He felt this could be made possible by using a suitable symbolic system for knowledge representation, which he called *characteristica universalis*. The computation system that he introduced with the name of *calculus ratiocinator* was aimed at extending algebraic methods to logic and thereby to proving or disproving any statement whatsoever. In a nondistant future, Leibniz thought, faced with a problem of any nature, instead of losing time in useless quarrels, *"wise and good willing men"* might sit around a table and say: *"Calculemus!"*; at that point they might run an algorithm (or even turn on a computing machine[8]) that would solve the problem.

The name of Leibniz is often related to the history of computers since he was the first to introduce the binary numbering system and the logical operations on which the binary mathematics used by computers is based. Actually, the relationship between Leibniz and computers is much deeper. We might say that a red line connects directly ideas and works of Leibniz across three centuries with the birth of the first computers.[9]

The construction of the first computers was, indeed, the result of a number of technologically and methodologically convergent developments that took place in the nineteenth and twentieth centuries and of the pressing needs for computing power deriving from the industrial development and (most important) from the military build-up in which the German, British, and American governments were involved from the late 1930s to 1945. This is not the place to illustrate the history of computing machines, but we have to remember that a crucial role in this history (in particular, in the history of the first programmable machines, the real ancestors of today's computers) was played by the studies and works of the English mathematician Alan M. Turing in the mid-1930s. Such studies were devoted, on one side, to formalizing the concept of algorithm and understanding its limits by showing the existence of problems that algorithms are unable to solve. On the other side, Turing's work suggested the possibility to build programmable computers, that is, computers that could execute whatever algorithms were assigned to them in a suitable formal description.

The connection between Leibniz and Turing in computer history, more properly in the history of algorithms, has been established thanks to a problem formulated by another great mathematician, David Hilbert. Among the 23 problems that Hilbert presented at the Mathematics World Congress, held in Paris in 1900, as the main open problems to which the work of mathematicians should be devoted in the

---

[8]In 1673 Leibniz himself proposed one of the first models of computing machine to the Royal Society: the *"machina arithmetica."*

[9]It is worth observing that the creation of suitable ontologies allowing us to classify entire domains of human knowledge is still today one of the big challenges of modern computer science. This is related to the aim of providing computer systems (not only those that supervise information search in the Web but also those that execute traditional data management applications) with semantic support that enhances the "intelligence" of software.

twentieth century, the Second Problem emerged as the most important. The problem consisted in establishing whether the axioms of the logical theory of arithmetic were consistent or not, or, in other words, whether it was not possible to derive two contradictory consequences from the axioms.

The issue was more precisely addressed in a famous talk given by Hilbert at the Mathematics World Congress that took place in Bologna in 1928. On that occasion, besides underlining the relevance of the issue of consistency, Hilbert illustrated two other problems concerning the logical theory of arithmetic: the completeness of the theory, that is, the property by which any true assertion should be provable in the theory, and the decision problem (the *Entscheidungsproblem*), that is, the problem of establishing whether there exists an algorithm that, given any logical formula of the theory of arithmetic, is able to tell, in a finite number of steps, whether the formula is valid or not. In other words, Hilbert proposed to address an issue that was much more specific than the ambitious one raised by Leibniz. In fact, the question was not to assess the existence of an algorithm able to decide the truth or falsity of any statement in any field of knowledge but was limited to considering the formalized knowledge in a specific logical domain and to finding out whether an algorithm existed able to decide truth or falsity for an assertion in such a theory. Clearly a positive reply to Hilbert's question would have been encouraging with respect to Leibniz' dream to decide in algorithmic terms any dispute, while a negative answer not only would have implied the end of this dream but would have also indicated a precise limit to the power of algorithmic methods. This type of issue, the conceptual tools needed for addressing them, and the answers that eventually were achieved, had a fundamental role in twentieth century mathematics and also had unforeseen consequences in the development of computer science and of its conceptual basis (see also Chap. 3).

The negative answer to the *Entscheidungsproblem* was given by Alan Turing in 1935–1936. How was this result obtained? In order to show that a problem can be solved by means of an algorithm, it is sufficient to show this algorithm. In this way we can prove the existence of an algorithm to compute the greatest common divisor or to decide whether an integer is a prime number or not. Instead, in order to show that a problem cannot be solved by means of an algorithm (or, as we used to say, to prove that it is an undecidable problem), first of all it is necessary to provide a definition of the concept of algorithm and then to show that algorithms corresponding to this definition cannot solve the given problem. Despite various previous attempts in this direction, in 1935 a satisfactory definition of the concept of algorithm was not known and, hence, first Turing had to find one.

To start with, Turing introduced a notion of algorithm that was based on a very elementary abstract machine model (later to be called "Turing machine"). Such a machine is provided with a (potentially) unlimited tape divided into cells, on which a tape-head (able to move in both directions) can write and read symbols belonging to a suitably defined finite alphabet. In addition, the machine in any moment is in one of a finite set of internal states. For example, a typical behavior rule of the machine could be the following: if the machine is in state q_1 and its head reads the character '2' on the tape, then it should write the character '4' in place of '2' in the same cell, move the head to the next cell to the right, and enter the state q_2. Note that, in general, a machine that is activated with a particular string x of characters (the "input" string) on the tape can reach a particular "final" state corresponding to the end of the computation. In this case the content of the tape in such a moment can be considered the "result" of the computation, but in some cases it may happen that the machine does not reach the final state and may indefinitely keep working.

The next step of Turing's work was the following: having assumed that the most general kind of algorithm was provided by his machines, he had to identify a problem that could not be solved by these machines, since a problem not solvable by means of Turing machines would be a problem that could not be solved by means of any possible kind of algorithm. Such a problem turned out to be exactly the so-called halting problem of Turing machines. In fact, Turing was able to show that there does not exist any Turing machine able to decide, given a machine M and a string of symbols x, whether M enters a final state or not when it is activated on a tape containing the input string x. Finally, as a last step, Turing showed that the halting problem could be formulated as a particular case of the decision problem stated by Hilbert.

As stated above, Turing reached several results that had a fundamental role in the subsequent developments leading to the construction of the first computers and that are among the most important conceptual pillars of computer science. The first result, clearly, was to provide a definition of the concept of algorithm that, although very simple, has never been substituted by any other definition and still represents the most general notion of algorithm. In fact, no problem is known that can be solved by any algorithm and that cannot be solved by a Turing machine.

The second result was to identify the limits of algorithms. After the proof of the undecidability of the halting problem (and, as a consequence, of the *Entscheidungsproblem*), many more undecidable problems have been discovered in various domains of knowledge (algebra, geometry, coding theory, computer science, etc.). The same limits that hold for Turing machines clearly hold for real computers in the sense that if a problem cannot be solved by means of algorithms, it cannot be solved by any computer, no matter how powerful it is.[10]

The third, fundamental result is closely related to the construction of programmable computers. Although Turing was not thinking in terms of real computing machines when he defined his model, but rather he was trying to represent the way a man's mind proceeds when he performs a computation, it is clear that his notion of algorithm (based on the Turing machine) was leading to the conclusion that any algorithm could be performed with an automatic device. But this might create the impression that we should build a different machine for every problem that we want to solve. The most surprising result that Turing showed

---

[10]Note that the contrary is not true. As we will see in Sect. 2.3 and in Chap. 3, there are problems that, in principle, can be solved in algorithmic terms but cannot be solved in practice with a computer since their solution may require an amount of time (computation steps) greater than the life of the Universe.

is that, instead, there exist special Turing machines able to simulate the behavior of any other Turing machine. A machine U of this kind, called a *universal Turing machine*, is devised in such a way that if we provide the description of another Turing machine M (the "program" of M) and a sequence of symbols x (the "input" of M) on the tape of U and then we activate U, the universal machine, step by step, executes the rules of the machine M on the string x.

The universal Turing machine had an important role in the invention of programmable computers that in the 1950s replaced the previously built computers, thanks to its greater versatility. A great deal separates those first programmable computers, in which the program had to be manually uploaded in the memory of the computer in the form of a binary string, from the modern computers in which programs are written in high-level languages easily understandable by the users and then translated into binary computer language by suitable software systems running on the computer itself. On the other hand it has been precisely this long history of innovations and technological advances that has made algorithms and computers ubiquitous today.

1.8 Conclusion

In this chapter we have shown, by means of a few paradigmatic examples, how in various historical ages, spanning thousands of years, algorithms were used to solve problems of interest for specific applications or, in some cases, simply to solve recreational problems. The examples also show a variety of styles in which algorithms were presented in different contexts. We have also illustrated how, around the mid-twentieth century, the need to provide a formal definition of the concept of algorithm and the discovery of the existence of universal algorithms (or machines), able to interpret and execute any algorithm presented to them (provided it is written in a suitable formal language) had a crucial role in the construction of the first programmable computers.

Today, thanks to the ubiquitous presence of computers and of computer applications, the concepts of algorithm and of computer program are quite familiar also to non-experts. Nonetheless, it is not as clear to everybody that, thanks to the technological development of computers and of programming languages and to the growing complexity of applications, the design of algorithms has changed from a creative and artisanal activity into a real science. In the next chapter we will show how the design of valid and efficient algorithms requires the use of suitable design techniques and a sophisticated organization of the data to be processed. Then, in the subsequent chapters, we will present examples of algorithms that are present in our everyday life (although hidden in devices such as, for example, our cellular phones) and we will illustrate the advanced design techniques that have been employed in order to obtain from them the required performance and effectiveness.

1.9 Bibliographic Notes

The history of algorithms and the history of mathematics clearly have several points in common. In order to understand such interconnections it is certainly advisable to read a volume devoted to the history of mathematics, such as [10], but, more interesting may be [35], where a study of the role that computing had in the evolution of mathematics is presented.

The history of algorithms is systematically presented in [16]. From this work are taken various examples presented in this chapter. Several volumes illustrate the development of the concept of algorithms in some specific historical periods. Particularly interesting are the books [15] for Egyptian mathematics and [95, 114] for Arab mathematics and, in particular, for the works of al-Khwarizmi. A good illustration of the relevance of the work of Fibonacci in the history of algorithms is contained in [28], while a general historical perspective is provided by Morelli and Tangheroni [81]. The Trémaux algorithm for the traversal of labyrinths is presented in [77].

The interesting relationships among mathematics, philosophy, and esoterism that characterized the search for a language that could be used to classify and represent all human knowledge is well described in [37]. The cultural and conceptual trail that, in the context of Western mathematical thought, leads from Leibniz to Turing and to the invention of programmable computers is illustrated in [23] in a very clear way by one of the greatest contemporary logicians, Martin Davis, who had an important role in the study of undecidability. Readers interested in the life and work of Alan M. Turing can read [60].

Chapter 2
How to Design an Algorithm

Rossella Petreschi

Abstract Designing an algorithm is a profoundly creative human endeavor. Indeed, to design an algorithm one has to conceive a solution by drawing on a deep understanding of the problem at hand, on one's knowledge of techniques adopted for the construction of other algorithms and, above all, on a fair sprinkling of one's personal inventiveness. As a consequence there can be no fully automated method for generating the solution to a given problem. So, in this chapter we want to indicate the line to be followed in order to arrive at an algorithm design of optimized form. Integral to this we explain why it is essential to find the best way to abstract, represent and organize the information available about the specific problem to be tackled.

2.1 Introduction

In the process leading to the creation of an algorithm to solve a given problem it is possible to identify two basic phases: first, outlining the mathematical kernel of the problem, and second, identification of appropriate techniques for designing the procedures that will lead to its solution. Obviously, these two phases are not clearly separable; on the contrary, they are tightly intertwined. Algorithm design, in fact, is not to be understood as solely a process for developing solutions to problems that are already well-defined, but also as a methodology for clearly defining the problem to be tackled in the first place. As a consequence, the design of algorithms can be seen as an art, for it can never be rendered fully automatic because it is founded on a thorough understanding of the problem to be solved, on analysis of approaches adopted for the construction of other algorithms, and, crucially, on individual insight and creativity.

R. Petreschi (✉)
Dipartimento di Informatica, Sapienza Università di Roma, via Salaria 113, 00198 Roma, Italy
e-mail: petreschi@di.uniroma1.it

In this chapter the design of a "good" efficient algorithm is presented; namely we first provide an introduction to the mathematical concept of a graph (Sect. 2.2), then we go on to illustrate two of the best-known techniques of algorithm design (Sect. 2.3). Following this, we introduce the concept of the *goodness* of an algorithm as a measure of its efficiency (Sect. 2.4). Finally, in Sect. 2.5, we present a complete example of the design of an algorithm, focusing on the *binary search algorithm.*

2.2 Graphs

Let us start by providing an introduction to the mathematical concept of a graph, which will also pave the way for several subsequent problems presented in this book where the search for an efficient solution entails the modeling of information in the form of graphs.

2.2.1 *The Pervasiveness of Graphs*

A graph is the representation of a set of elements where some pairs of the elements are connected by links. By convention, the elements of a graph are called nodes and are represented as points. The link (relationship) between two nodes is represented by a line. Depending on whether the relationship is symmetrical or non-symmetrical, a graph is defined as non-orientated or orientated (Fig. 2.1). In other words, in a non-orientated graph the relations (x, y) and (y, x) correspond to the same line, called an edge, whereas in an orientated graph (x, y) and (y, x) refer to two different lines, called arcs.

A graph is a natural way to model a real-life situation because it represents entities in relation to one another. As an example, one can think of a railway network: the stations are the nodes of a graph whose edges indicate that two stations are connected by a section of the railway track. Clearly, networks of any kind can be represented by graphs. In this book we focus on three types of network. Two of these, the television network (Chap. 8) and the Internet (Chap. 5), concern systems that exchange messages in electronic form through links that make use of cables, fibre optics, radio or infrared connections; the third is the road network (Chap. 4).

If it is easy to represent a network in terms of graphs, it is not so natural to think of graphs as capable of representing labyrinths (Sect. 1.6), or a sporting activity, or even as a useful aid in the search for the ideal marriage partner. Nevertheless, a sports tournament, for example, can be modeled with a graph, each node representing a participating team and two teams being linked once they have played their match against each other.

As for the search for a marriage partner, *the problem of stable marriages* can be modeled as follows. Suppose that in a small, isolated village there are equal numbers of young men and women of marriageable age. Given the village's isolation,

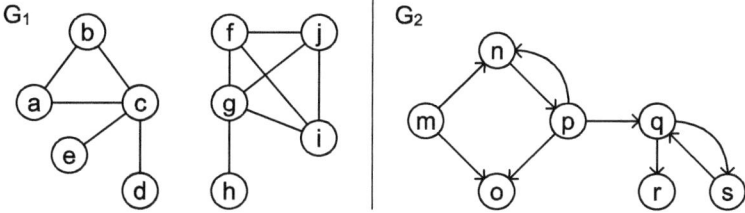

Fig. 2.1 Examples of graphs: G_1 non-orientated graph; G_2 orientated graph: the *arrows* on each arc indicate the orientation of that arc; for example, the arc (q, r) exists, but the arc (r, q) does not

Fig. 2.2 Affective ties represented through graphs. *Solid lines* represent possible matches. (**a**) A graph representing a state in which at most three marriages are possible. (**b**) A graph showing a situation in which four marriages can be celebrated

its elders naturally hope that the maximum number of these young people will be able to find a marriage partner among the pool of available mates, in order that reproduction of the population can be ensured. Unfortunately, however, it is not certain that all the eligible young people will succeed in "tying the knot", because obstacles such as too-close kinship ties, or insurmountable dislike for someone or other, will inevitably exclude the possibility of some marriages.

In order to establish the maximum number of couples that can be formed, it turns out to be useful to model the problem by way of a graph of $2n$ nodes, n representing the young women and n representing the young men. The edges in the resulting graph have to provide information on the possibility of forming a marriage that can produce offspring; thus they will only connect male nodes with female nodes. An edge will be present only if there is no impediment to the marriage.

Figure 2.2 shows two possible states of affective ties between eight young people of the two sexes. The ideal would be to end up with four marriages, so as to ensure the prosperity of the village. But observing the relations in the first case, shown in Fig. 2.2a, it is apparent that this is not possible. In fact, Julia can only marry Alexis, just as Clair can only marry Daniel. Alexandra and Benny, instead, each have two possible matches, Alexis and Daniel and Gabriel and Mark, respectively.[1] The obligatory matches of Julia and Clair therefore leave Alexandra without a partner, whereas any match made by Alexandra would leave either Julia or Clair without a partner. The consequence of Benny's choice will be that either Gabriel or Mark will

---

[1] The graph could, instead, be read symmetrically from the male point of view, that is, Alexis can choose between Julia and Alexandra, Daniel between Clair and Alexandra, etc.

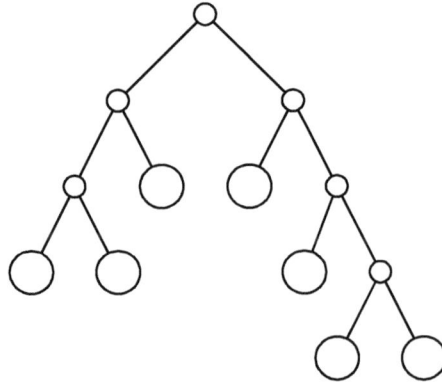

Fig. 2.3 The binary tree associated to the mobile in Fig. 2.4

remain bachelors. The other situation, shown in Fig. 2.2b, is clearly more favorable. Since in this scenario Clair has no incompatibility with any of the four young men, it follows that four marriages can be celebrated.

A wide variety of algorithms have been designed for solving this problem. Each of these algorithms, given a graph representing a particular instance of the problem, enable it to be decided whether or not it is going to be possible to form couples from all members of a given group of young people, and if it is not, to establish the maximum numbers of couples that it is going to be possible to form.

It should be noted that this problem, which has been presented in a playful way here, is, in fact, of great practical interest. To give just one example of its many potential applications we mention the problem of assigning the staff of a company to particular shifts.

An important class of graphs which offers a wealth of potential applications is that of *trees* (box "Trees").

Here, let us introduce trees by comparing them to the moving abstract sculptures known as "mobiles", thereby further emphasizing the connection between scientific and artistic disciplines already highlighted in Chap. 1.

Alexander Calder (1898–1976) revolutionized the world of sculpture by introducing movement as a fundamental component of his works. His mobiles, today available in series and sold in shops of the world's leading art museums, pivot on just one point and are free to move slightly according to air currents in the environment in which they are located. Beyond their artistic value, these sculptures are perfect examples of binary trees. In fact, each joint of a mobile has exactly two terminal points, each of which can be either just a terminal, or a new joint which has, in turn, two terminal points. Figure 2.3 depicts the tree associated to the mobile in Fig. 2.4, where it can be seen that the root corresponds to the pivot.

Note that we could attribute labels 0 or 1 to the arcs coming out of a node, according to whether they lead to a leaf or to another internal node. Each sequence of zeroes and ones, corresponding to the path from a root to a specific leaf, enables information to be given regarding the position of that leaf in the tree. In other words, the sequence of zeros and ones is a variable-length code that is representative of that

Fig. 2.4 Mobile in the garden of the Beyeler Foundation in Basel (photo Baglioni)

leaf in that tree. Note that this concept is very similar to the one used for the Morse code (1838), which combined sequences of variable length lines of dots, dashes and spaces to represent the letters of the English alphabet. These simple considerations could initiate a long discussion on how to construct an efficient binary code for a computer, but that is another story and the interested reader will find useful references in the bibliography.

Trees

Given a graph $G = (V, E)$ with n nodes $V = (v_1, v_2, \ldots, v_n)$ and m edges (v_i, v_j), we call *path of length* k from a node v_i to a node v_{i+k} a sequence of nodes $v_i, v_{i+1}, \ldots, v_{i+k}$ such that (v_{i+j}, v_{i+j+1}) is an edge of G, with $0 \leq j \leq k - 1$. When $v_i = v_{i+k}$ the path is a *cycle*. A non-orientated graph G is said to be *connected* if every pair of nodes v_i and v_k is joined by a path.

A *free tree* T (T_1 in Fig. 2.5) is a connected non-orientated graph without cycles. A tree with n nodes has $n - 1$ edges and for any pair of nodes there exists just one path that connects them. A *rooted tree* R (T_2 in Fig. 2.5) is a tree in which a particular node, defined *root*, has been identified. In a rooted tree there exists an implicit direction between the root and the other nodes that allows the following definitions to be introduced:

- w is the *father* of v; that is, v is the *son* of w if there exists the edge (w, v) on the path from the root to v;
- f is a *leaf* if it does not have children. In a tree a node is either a leaf, or it is an *internal* node;
- The *height* of a tree is the longest path among all those leading from a root to a leaf.

(continued)

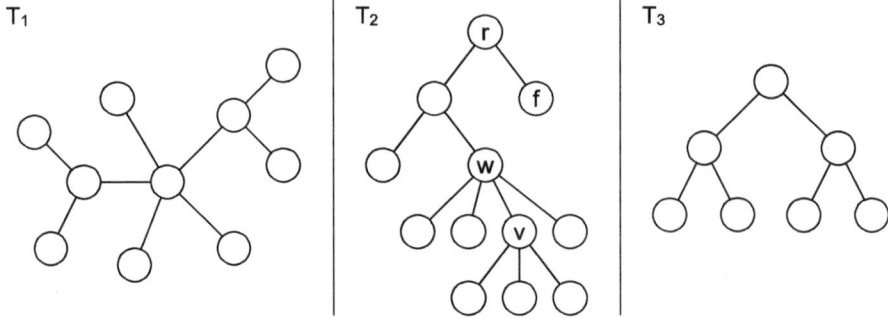

Fig. 2.5 Examples of trees: T_1 free tree; T_2 rooted tree; T_3 complete binary tree

> **(continued)**
>
> A rooted tree is said to be a *binary* tree if each node has at most two children; furthermore, a distinction is always made between *left child* and *right child*, even if one of the two children does not exist. A tree (and a graph, in general) is said to be *weighted* when a cost is assigned to each of its edges. A tree in which every internal node has exactly two children is called a *complete* binary tree (T_3 in Fig. 2.5) and a binary tree in which every internal node has exactly one child is called a *degenerate* binary tree. The height h of any given binary tree is a value falling somewhere in the interval between $\log_2 n$ (height of a complete binary tree having n nodes) and n (height of a degenerate binary tree having n nodes).

2.2.2 The Origin of Graph Theory

Graph theory is one of the few fields of mathematics which has a precise date of origin. The first work on graphs was written by Euler in 1736 and appeared among the publications of the Academy of Science in St. Petersburg. Euler had found himself presented with the following problem:

> In the city of Königsberg[2] there is an island, A, called Kneiphof, around which flow two branches of the River Pregel. Crossing these two branches are seven bridges: a, b, c, d, e, f, g. The question is this: is it possible for a person to choose a walk which takes them across each of the seven bridges once and only once, and then brings them back to their point of departure?

---

[2]Today named Kaliningrad.

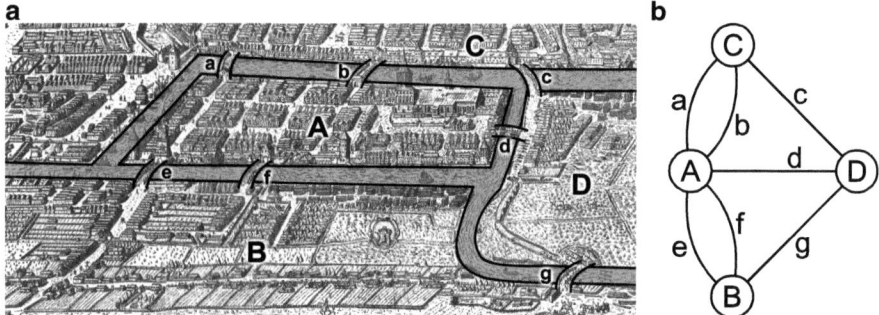

Fig. 2.6 (a) The city of Königsberg and its bridges. (b) Representation of city of Königsberg's map in graph form

Figure 2.6a shows a map of the city of Königsberg with the four districts which Euler labelled A, B, C and D, and represented as the nodes of a graph connected by edges corresponding to the bridges (Fig. 2.6b).[3] Euler proved that the graph can not be visited completely following a single cyclic path; that is, regardless of which node is departed from, it is impossible to visit all the nodes of the graph and return to the starting node without backtracking for at least part of the way.

A path solving the problem (today called an *Eulerian circuit*) would enter each node exactly the same number of times as leaving it, but this condition cannot be satisfied by the map of Königsberg. Euler subsequently went on to prove that every connected graph with all nodes of even *degree*[4] exhibits an Eulerian circuit.

From the foregoing description, it is clear that an Eulerian circuit requires that each arc of a graph be visited once, and only once. When, instead, it is required that each node of a graph be passed through once, and only once, this is known as a *Hamiltonian Cycle*.

In 1859 the Irish mathematician W.R. Hamilton created a wooden puzzle in the form of a tridimensional regular dodecahedron (Fig. 2.7a) (subsequently distributed commercially as a pegboard with holes at the nodes of the dodecahedral graph). The name of a city was associated to each of the 20 nodes of the dodecahedron and the problem was to find an itinerary along the arcs such that each city was visited once and only once. The path selected by a player was recorded by a thread wound round pegs inserted into the nodes.

Since this puzzle did not have much success on the market, Hamilton brought out a two-dimensional version that was easier to handle: the puzzle was in the form of a *planar* graph[5] corresponding to a dodecahedron (Fig. 2.7b). However, it seems

---

[3]To be precise, the graph in Fig. 2.6b is in fact a multigraph, since two of its nodes can be connected by more than one edge.

[4]The degree of a vertex in a non-orientated graph is defined as the number of edges incident on that vertex.

[5]A graph is said to be planar if it can be drawn in the plane without its edges intersecting.

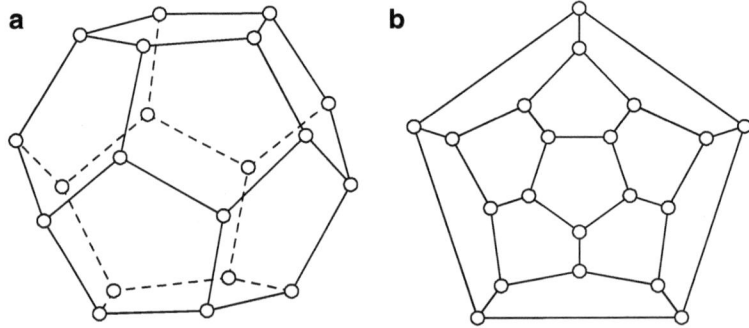

Fig. 2.7 (**a**) A tridimensional regular dodecahedron and (**b**) its representation as planar graph

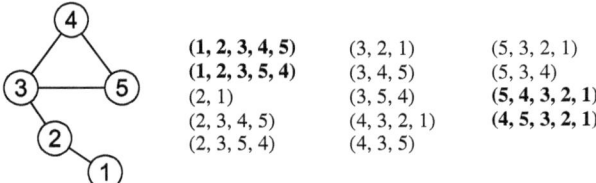

Fig. 2.8 List of the 14 paths on a simple graph of 5 nodes that do not pass more than once through the same node. The four paths of length 5 are Hamiltonian

that this puzzle did not meet with much success either. In contrast, the search for a Hamiltonian Cycle has, from that time onwards, been considered one of the most important problems of graph theory.

At this point we need to pause a moment to explain that, although the calculation of an Eulerian Circuit and a Hamiltonian Cycle are very similar problems as regards their formulation, they are very different as regards their degree of difficulty. First, let us point out that the solution to a problem modeled in terms of a graph is obtained by analyzing whether or not a structural property of that graph satisfying the specific instance of the problem exists. Now, as regards the Eulerian Circuit we know (indeed, from Euler himself) that a connected graph has an Eulerian Circuit if all the degrees of its nodes are even. Since establishing whether the degrees of a graph are even is easy, it follows that it is also easy to establish whether a graph has an Eulerian Cycle. However, the situation is quite different when it comes to Hamiltonian Cycles.

To date no structural property of the graph associated with the Hamiltonian Cycle problem has been found, and it is not known if one exists. As a consequence, whether or not a Hamiltonian Cycle exists can only be determined by exhaustive analysis of all the possible itineraries along the arcs of a graph. In fact, identifying all these possible itineraries is in itself a difficult problem, as becomes clear from a quick glance at the number and complexity of itineraries existing in even a small graph, such as the one shown in Fig. 2.8. This explains why, even today, finding a Hamiltonian path remains "a difficult problem to solve" (see Sect. 2.4).

2.2.3 The Topological Ordering Problem

Consider the following problem: a part-time student, studying for a university degree in Informatics in Rome, has to choose which courses to follow among those she is expected to take. Owing to her work commitments, our student can only take one course in a semester; the courses can be followed in any order, but only as long as the prerequisites established by the Faculty Academic Board are satisfied.

Thus, owing to the fact that an acceptable choice exists, it is clear that:

C1: *The prerequisites must be such that they are not mutually binding: if course A is a requirement for proceeding to course B, and course B is a requirement for proceeding to course C, then course C cannot be a requirement for proceeding to course A, otherwise no acceptable order for the courses would exist;*

C2: *At least one of the courses must not have an entry requirement. It must be stressed that in general there will not be just one possible selection.*

Let us see how our student goes about making her particular selection. As a starting point she decides to represent the courses by means of the nodes of an orientated graph and the prerequisites by arcs between these nodes: the arc from node v to node w indicates that course v has to be completed before starting course w. The problem of the choice of courses, when represented on an orientated graph, is the equivalent of asking oneself if an ordered sequence of nodes exists such that v precedes w, for every arc (v, w) on the graph, while the conditions C1 and C2, expressed on the graph, become:

G1: *No cycles must exist. If we analyze a cycle of three nodes a,b,c and three arcs (a, b), (b, c), (c, a) it becomes apparent that none of the six possible sequences of three nodes (a, b, c), (a, c, b), (b, a, c), (b, c, a), (c, a, b), (c, b, a) is an acceptable solution, given that for each sequence there exists at least one arc whose end points are not in the correct relative order in the sequence (the arc (c, a) in the first sequence, the arc (b, c) in the second, and so on);*

G2: *There has to be a* source, *that is, there has to be at least one node to which no arc arrives.*

At this point, the student is ready to make her choice by proceeding source by source. Let us suppose that the graph in Fig. 2.9a is the model she has constructed. A, B, C, D, E, F and G are the names of the courses and each arc XY indicates that course Y has course X as its prerequisite (e.g., course A is a prerequisite for courses B and D). We can follow the student's choices by observing the graph in Fig. 2.9a. This graph satisfies both condition G1 (there are no cycles) and condition G2 (there is at least one source[6]).

Let us now suppose that C is the first course that the student decides to follow. Once she has sat and passed the exam for course C, both C and all the arcs leaving it are eliminated from the graph. This is equivalent to saying that in mastering the

---

[6]In the graph shown in Fig. 2.9a there are two sources: node A and node C.

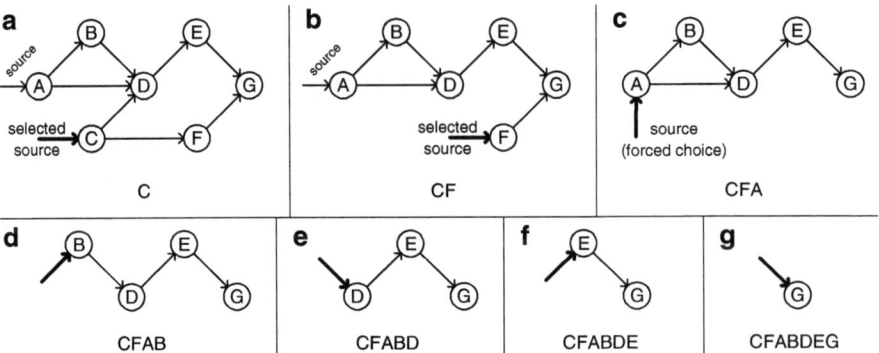

Fig. 2.9 Example showing how the topological ordering algorithm works

subject matter of course C she now has the necessary competences to tackle the subject matter of courses D and F. The new graph is shown in Fig. 2.9b. Since conditions G1 and G2 are still satisfied she can proceed with a new choice (A or F). Let us say she chooses F. The elimination of F and its outgoing arcs leads to a graph that has just one source. That is, she has become obliged to tackle course A before being able to take any other course (Fig. 2.9c). The rest of Fig. 2.9 shows the remaining steps of the procedure, and we can see that, in the case of this example, from this point onwards the choice of course is always obligatory and the gradually reduced graph obtained at each successive step satisfies both G1 and G2.

The sequence is completed when all the nodes and arcs of the graph have been eliminated. It must be emphasized that the solution CFABDEG found by the algorithm here is not the only one possible. A quick look at Fig. 2.9a shows that A could have been chosen as the first source and thus as the first element in a differently ordered sequence.

2.3 Algorithmic Techniques

In the previous section, having modeled the problem of topological ordering in terms of graph theory and identified the properties of the model graph (properties G1 and G2), we went on to present the appropriate algorithmic technique for resolving the given problem. In that case it was the procedure of successively analyzing each reduced graph obtained by eliminating a specific source.

It is by no means easy to explain how one arrives at the design technique to be employed in constructing an algorithm because, as we said at the outset of this chapter, the choice of technique depends not only on a thorough understanding of the problem, but also on the extent of our knowledge about techniques that have already proved effective in solving other problems. It also depends upon a good dose of intuition in hitting upon the model that is most suited to the problem at hand.

In this section we introduce two of the most interesting and widely used algorithmic techniques and illustrate them by way of problems drawn from everyday life. These are the *backtrack technique* and the *greedy technique*.

In general terms we can say that the backtrack technique is applied when nothing is known about the problem to be solved apart from the statement of the problem itself. One thus proceeds by making choices that can be reversed if needed. The greedy technique, instead, is applied when one has sufficient information about the problem to be able to proceed by making local irreversible choices.

In the final section of this chapter a further highly useful algorithmic technique will also be introduced, that of *divide et impera (divide and conquer)* which can be applied when a problem lends itself to being divided into various subproblems.

2.3.1 The Backtrack Technique

The *Eight Queens Puzzle*, in its original form, is the problem of finding how to place eight chess queens on a chessboard (8×8 squares) in such a way that no two queens attack each other.

For readers who are not familiar with the rules of chess, it is sufficient to know that a queen can move horizontally, vertically or diagonally for any number of squares; consequently, a solution requires that no two queens share the same row, column or diagonal.

The first proposal of the puzzle was due to Max Bezzel, who in 1848 published it in the chess journal *Berliner Schachzeitung*. The problem was considered so intriguing that even Carl Friedrich Gauss devoted time to studying it. In 1874 Professors Glaisher of Cambridge University and Günther of Leipzig University published the first proof that there are 92 solutions to the problem. These 92 solutions are obtained from 12 fundamental solutions by rotating the chessboard 90°, 180°, or 270° and then reflecting each of the four rotational variants (including its original form) in a mirror in a fixed position.

Figure 2.10a shows the 12 fundamental solutions to the eight queens problem. In each sequence of eight numbers the order in the sequence indicates the row—and the number itself, the column—in which each of the eight queens is placed. Figure 2.10b shows in detail the situation represented by the sequence 1 5 8 6 3 7 2 4, with the identifiers Q_1 to Q_8 representing each queen (and corresponding to the number of the row she occupies). Thus, the first queen is placed in the square at the top left of the board (first row, first column), the second queen is placed in the second row, fifth column, and so on. Note that in order to represent a possible positioning of the eight queens by way of a sequence, just eight items of information are sufficient. In contrast, in order to represent the positioning by way of a chessboard we need to know the content of all 64 squares (see box "Data Structures").

The first solutions to the eight queens problem were based on purely numerical calculations. Here, however, we present an algorithmic solution that is arrived at through an exhaustive, but controlled, analysis of all possible positions that each

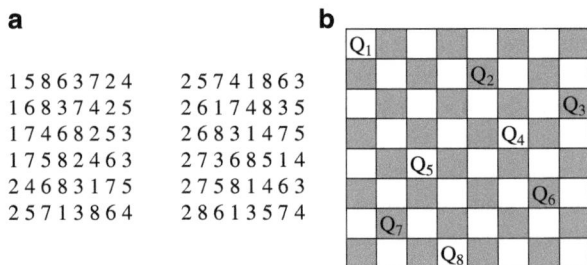

Fig. 2.10 (**a**) The 12 fundamental solutions to the eight queens puzzle. (**b**) Chessboard showing the positioning of the eight queens represented by the sequence 1 5 8 6 3 7 2 4

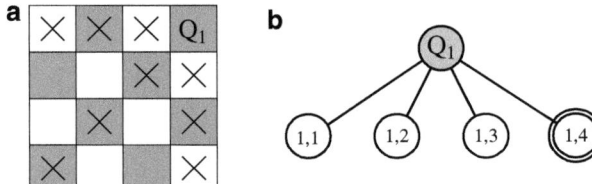

Fig. 2.11 Placing of the first queen in the last column of the first row: (**a**) Represented on a chessboard; (**b**) represented on a decision tree

successively placed queen can occupy. Given the conditions that each placement must comply with, a systematic search is designed which, by means of checks, eliminates a certain number of positions in advance, thus reducing the number of possible placings that need to be analyzed. To make the illustration simpler, let us consider the problem of placing four queens (Q_1, Q_2, Q_3, and Q_4) on a 4×4 chessboard. At the start of the algorithm, Q_1 can be placed on any one of the four columns of the board. Let us choose the fourth column—that is, let us put 4 as the first number in the solution sequence.

This choice, as can be seen from Fig. 2.11a, means the other three queens are unable to use the first row, the fourth column and the diagonal towards the right i.e., the one linking square $(4, 1)$ with square $(1, 4)$. Figure 2.11b shows the first step in the construction of the *decision tree*,[7] a tree that will enable us to represent the procedure that we are carrying out. Each shaded node contains the name of the queen that is being allocated to it, while all the unshaded nodes (that are children of a shaded node) indicate possible alternative positions for that queen, taking into account the choices made in all preceding placings. Since it is the first queen that we are placing, all four positions on the board are still available. The double circle in an unshaded node (see Fig. 2.11b) indicates that it is the position that has currently been chosen for the queen that is being analyzed.

---

[7]Decision tree is the name given to a tree in which each node represents a set of possible choices.

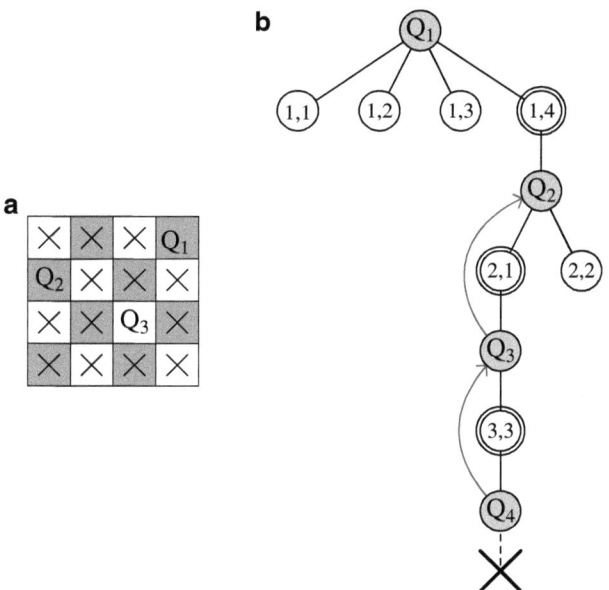

Fig. 2.12 Impossibility of placing the fourth queen following the choice to place the first queen in the fourth column and the second queen in the first column. (**a**) Representation on a chessboard; (**b**) representation on a decision tree

The second queen, therefore, can only be placed on either the first or the second position on the second row, given that the third and fourth positions are, respectively, on the diagonal and on the column of Q_1. Let us begin by choosing the first position, that is, by writing 1 as the second number of the solution that we are seeking. As a consequence of the previous choices, Q_3 can only be placed on the third position of the third row, that is, 3 is the third number of our hypothetical solution, given that the first position is on the column of Q_2, and the second and fourth on the diagonal and column, respectively, of Q_1. As a consequence of the choices made, Q_4 does not have any acceptable placing, given that the first, third and fourth squares are on the columns of Q_2, Q_3 and Q_1, respectively, while the second square is on the diagonal of Q_3 (see Fig. 2.12a).

Given the impossibility of placing the fourth queen, the algorithm goes back to the last choice made in order to check if it is possible to arrive at a solution by another route. In our example the last choice made was that regarding the placing of Q_2, since the placing of Q_3 had turned out to be obligatory. Figure 2.12b gives the decision tree of the procedure up to this step, showing the allocation of queens Q_1, Q_2, and Q_3 and the backtrack due to the impossibility of allocating queen Q_4.

Returning to the shaded node Q_2, it becomes apparent that only one alternative placement possibility exists: allocating Q_2 to the second square of the second row. The choice to place Q_1 on the fourth square and Q_2 on the second square block the placing of Q_3, since the first and third squares of the third row are on the diagonal

Fig. 2.13 Impossibility of placing the third queen following the choice to place the first queen in the fourth column, and the second queen in the second column.
(**a**) Representation on a chessboard;
(**b**) representation on a decision tree showing the backtrack to the root—that is, the placing of the first queen

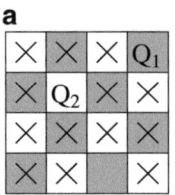
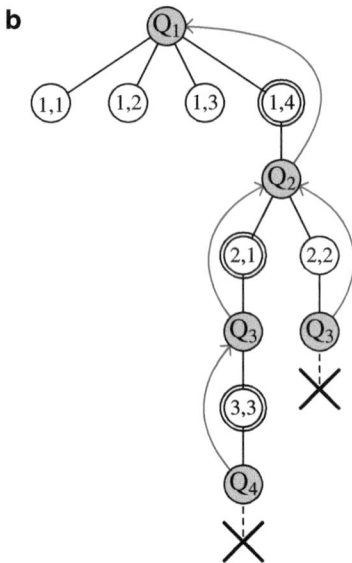

of Q_2, while the second and the fourth are, respectively, on the diagonal and column of Q_1 (see Fig. 2.13). Backtracking to Q_2 has shown us that there are no alternative choices for queen Q_3 under the hypothesis of Q_1 in position 4. Therefore a new search for a solution can only start from a different placing of Q_1 (Fig. 2.13b). Figure 2.14 shows the decision tree depicting the behavior of the entire algorithm, according to which of the four possible placements of Q_1 is chosen.

Data Structures

We have seen that an algorithm is a sequence of unambiguous instructions that enable an output to be generated in a finite time interval for every legitimate input.

The information that an algorithm receives as input has to be a meaningful abstraction of the actual data to be processed, that is an abstraction capable of representing the nucleus of information of the problem, and from which it is possible to derive the desired results.

In order to clarify this concept, let us take the problem of placing n queens on an $n \times n$ chessboard: the meaningful part of the data is an object possessing the property of being able to attack the row, column and diagonal associated with its position on an 8×8 matrix (precisely the characteristics of a chess queen). No other real-world information, such as, for example, the material from which the chess pieces are made, the precise shape of the pieces, or the craftsperson who made them, is necessary for solving the problem. Once

(continued)

Fig. 2.14 Decision tree showing all the choices consequent upon the four different placements of Q_1

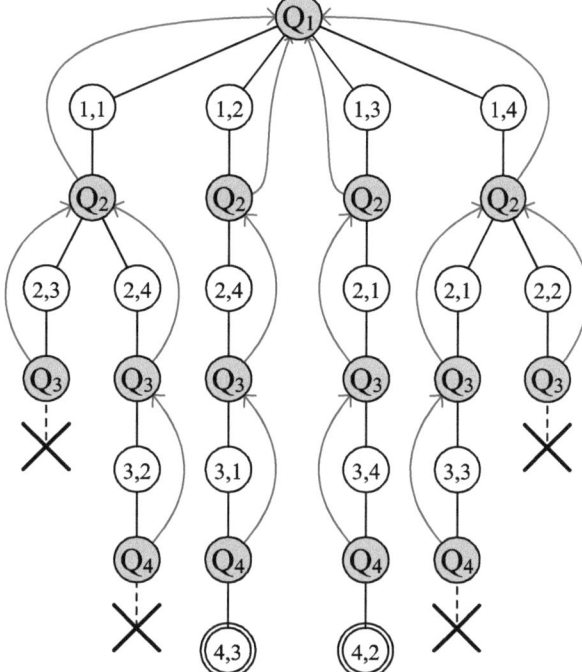

(continued)

the characteristics of the data to be given as input have been identified, the question has to be tackled of how to represent and organize them such that the algorithm is able to process them in the most efficient manner possible.

Referring once again to the problem of the queens, we noted that the representation by way of number sequences requires only n items of information in order to provide possible allocations for n queens in output, whereas the representation by means of a chessboard requires $n \times n$ items of information.

The choice of how to represent the data is, in general, anything but simple, and as the example of the eight queens has shown us, there is likely to be more than one mode of representation available.

Thus the choice has to be made with due regard to the type of operations that are going to need to be performed on the data. In our example the operations are those involved in checking the position to which a queen is allocated so as to ensure she is not attacked by any other queen already placed on the board. Using the representation by way of a matrix, if we wanted to place the ith queen on the square where the ith row and the jth column intersect we would first have to check that no other queen was already placed on the ith row, on

(continued)

(continued)

the jth column, on the diagonal to the left, on the diagonal to the right, up to a total of four checks before each and every hypothesis of placement.

Alternatively, how could we go about making these same checks using the representation by means of sequences? Here we present the solution proposed by Niklaus Wirth in his book *Algorithms + Data Structures = Programs*. Before placing the ith queen on the jth column, Wirth analyzes four binary variables which carry information about whether or not a queen is present in the ith row, r, jth column, c, in the diagonal to the right, rd, or in the diagonal to the left, ld. The variables can only assume two values, *true* or *false*, representing, respectively, the presence or absence of a queen. The $4n$ variables are all initialized with the value *false*, or *absence of a queen*, and then as each queen takes up her allotted place the relevant values switch accordingly. For example, placing queen 1 on position 4 leads to $r(1)$, $c(4)$, $rd(5)$ and $ld(-3)$ becoming *true*. The values 5 and -3 for rd and ld are two constants, derived from the fact that for rd the sum of the indices of row and column is always 5, while for ld the difference between these indices is always -3, and that these values, therefore, can be considered as representative of the two diagonals. In the case of a backtrack, the values corresponding to the removal of a queen from her previous placement will be reswitched to false. Thus it follows that adopting Wirth's approach it is again sufficient to make just four checks prior to each placement proposal. It is easy to see that this type of check is also efficient when the representation is done using a matrix, however in this case the disadvantage remains that of requiring a quadratic number of items of information.

In general, the term data structure refers to the way of representing and organizing the information an algorithm needs in order to generate a solution to the given problem. The choice of an efficient data structure is in itself a problem that can profoundly influence the planning and construction of a good algorithm. Queues, piles, lists, priority queues, search trees, heaps, tries, graphs, etc. are just a few of the many structures widely dealt with in the literature. The reader interested in exploring the subject further will find references in the bibliography at the end of this chapter.

2.3.2 The Greedy Technique

Let us now use as an example the simple act of giving a customer the correct change that cashiers all over the world perform daily. Typically a cashier will first choose, more or less automatically, the note or coin with the largest face value below the amount that needs to be given back, and then repeat this procedure until the operation is completed. For example, when giving change of 33 euro-cents the cashier will tend to start with a 20-cent piece, then a 10-cent piece, then a 2 and finally a 1. In this particular case, our mode of operation has led us to give the change using the lowest possible number of coins.

The algorithmic technique that we are now going to examine is known as the *greedy technique*, because at each step the aim is to grab the largest amount possible as a vital strategy towards achieving the final goal.

Let us analyze two further examples. Suppose the coinage we have is not euro-cents, but simply values of 6, 5 and 1 units (u). If change of $15u$ has to be given, keeping strictly to our greedy technique we would begin with a coin of $6u$, followed by another of $6u$ and finally, in sequence, three coins of $1u$, for a total of five coins overall. If we had been less greedy, the problem could have been solved with just three $5u$ coins and, as in our previous example, above, the lowest possible number of coins would have been used. Now let us suppose, instead, that the coinage available consisted of values of 7, 4 and 2 units. Greediness would lead us to choose two successive coins of $7u$. But then we would have to come to a halt, because any further coin would lead to the amount of change needing to be given being exceeded. So, in the first of these two examples the optimal solution of giving the change using the smallest number of coins possible was not achieved, while in the second, the procedure had to come to a halt without having produced a solution.

The obvious objections the reader may raise at this point are, first, that no cashier would be obliged to give change using the lowest possible number of coins, and, second, that as soon as a cashier realized that it was not going to be possible to arrive at the correct change, he/she would simply ask for the $7u$ back and then give the customer one $4u$ and two $2us$.

But what if the cashier were a machine? Every choice made would be irrevocable. What the greedy technique requires, in fact, is:

1. At each and every step an irrevocable choice be made which meets the demands of the problem (that is, to give an amount of change which does not exceed the amount due to the customer);
2. To pay no heed to what will happen afterwards but to be optimum locally (that is, to choose the coin with the largest face value among those that comply with rule 1).

The foregoing examples have highlighted the fact that it is necessary to ask oneself whether or not the choice of local optima will lead to a global optimum for the problem as a whole, given that there are certain problems for which the choice of local optima will lead invariably to an optimal global solution, and other problems for which this is not the case.

In fact, the optimality of a solution depends on the mathematical properties of the problem, or, to put it another way, it is necessary to be able to provide the mathematical characterization of the optimal solution that is being sought.

Our illustration will now go on to show that every greedy choice reduces the given problem to a problem that is of the same type, but of smaller dimensions. To put it another way, we have to prove that an optimal overall solution of the problem contains within it the optimal solutions of the subproblems of the same type. Finally, it is necessary to demonstrate by induction (see box "Induction") that the choices can be iterated.

Induction

In his "Formulaire de mathematiques" (1898), Giuseppe Peano showed that the theory of natural numbers follows from three primitive notions and five axioms. Thus:

Given 0, a number and its successor, and accepting that the natural numbers form a class, the following axioms are valid:

- *0 is a natural number;*
- *Every natural number n has a natural number successor (succ(n));*
- *Let S be a class of natural numbers containing zero. If every successor of every natural number in S still belongs to S, then S is the class of all natural numbers;*
- *Successors of equal natural numbers are themselves equal natural numbers;*
- *No successor of a natural number can be 0.*

Peano's system enables the following principle to be formulated:

Principle of Mathematical Induction *A statement is true, for every natural number n, if the following conditions are satisfied:*

Basic Case: *The statement is true for $n = n_0$, where n_0 represents the minimum of the indices that satisfy the statement (usually 0 or 1).*

Inductive Step: *The truth of the term for a generic natural number $n = k$ (inductive hypothesis) implies its truth also for the successive natural number $n = k + 1$.*

The term *mathematical induction* was coined by Augustus De Morgan in 1838, even if traces of the concept were evident prior to this date (to support this claim we cite "Arithmeticorum libri duo" (1575) by the Italian Francesco Maurolico as just one possible source). In the formulation of the Principle of Induction that has just been given, the term *statement* was purposefully generic because, by way of this principle, whenever we have an enumerable infinity of cases of uniform type with respect to the identifying index, it is possible both to formulate a definition and to provide a proof.

By way of illustration, we now give two definitions based on mathematical induction.

(continued)

(continued)

Definition 1. The sum $a + b$ of two natural numbers a and b is defined as follows: $a + 1 = succ(a); a + (b + 1) = (a + b) + 1 = succ(a + b)$.

Peano added an explanatory note to this definition affirming that it must be read as follows: *if a and b are numbers, and if $a + b$ is a number (that is, if it has been defined), but $a + (b + 1)$ has not yet been defined, then $a + (b + 1)$ means the number that follows $a + b$.*

Definition 2. $F_0 = 0; F_1 = 1; F_n = F_{n-1} + F_{n-2}$ for every $n \geq 2$

Note that this definition uses a generalization of the principle of induction in that it requires two basic cases, one for $n = 0$ and one for $n = 1$, and the inductive step requires two inductive hypotheses, one regarding F_{n-1} and one regarding F_{n-2}. This generalization goes by the name of *principle of complete induction* (the interested reader will find further references on this topic in the bibliography).

Proof by Means of Mathematical Induction. In order to prove a theorem T by mathematical induction (or, simply, by induction) it is necessary to prove that a statement E of T satisfies the two conditions of the principle of induction. As an example, let us attempt to prove that the sum of the first n odd numbers is n squared (Fig. 2.15), or to put it in another way, let us prove that: $1 + 3 + 5 + \ldots + (2n - 1) = n^2$.

The *base case* is trivially true.

The *inductive step* tells us that we must prove the truth of the assertion (a) $1 + 3 + 5 + \ldots + (2(k + 1) - 1) = 1 + 3 + 5 + \ldots + (2k + 1) = (k + 1)^2$, using the inductive hypothesis: (b) $1 + 3 + 5 + \ldots + (2k - 1) = k^2$.

Since $(k + 1)^2 = k^2 + 2k + 1$, the assertion (a) is obtained from the equality (b) simply summing $(2k + 1)$ to both sides of the equality:

$$1 + 3 + 5 + \ldots + (2k - 1) + (2k + 1) = k^2 + (2k + 1) = (k + 1)^2.$$

Now let us show that in order to complete a proof by induction it is essential to establish the truth of both the inductive hypothesis and the inductive step. In other words, let us show how mathematical induction enables us to avoid arriving at a result that is in error. Let us hypothesize that someone (erroneously) wishes to prove that $1 + 3 + 5 + \ldots + (2n - 1) = n$.

The *base case* for $n = 1$ is true. We have to prove that $1 + 3 + 5 + \ldots + (2k + 1) = (k + 1)$, assuming the truth of the inductive hypothesis $(1 + 3 + 5 + \ldots + (2k - 1) = k)$.

(continued)

Fig. 2.15 The sum of the first n odd numbers is n squared

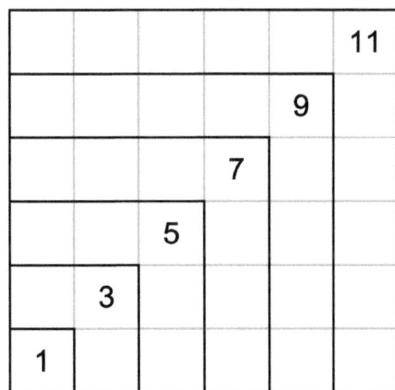

(continued)

Now, $1 + 3 + 5 + \ldots + (2k + 1) = (1 + 3 + 5 + \ldots + (2k - 1)) + (2k + 1) = k + (2k + 1) = 3k + 1 \neq k + 1$. Therefore it is not true that $1 + 3 + 5 + \ldots + (2k + 1) = (k + 1)$, given that the inductive step of the proof has led us to a different result.

Similarly, if we wished to prove that $1 + 3 + 5 + \ldots + (2k + 1) = k + 1$, the base case of the proof by induction would immediately make us realize the assertion was false, in as much as for $n = 1$ we would have $1 = 2$.

As an example of the type of problem for which the greedy technique provides optimal solutions, let us consider the generation of a *minimum spanning tree*.

Let G be a connected, non-orientated graph $G = (V, E)$; a tree T is called a *spanning tree* for G if it has the same n nodes as V, but only $n - 1$ of the m edges of E. To aid understanding of this concept we might say that a spanning tree represents a possible "skeleton" for the graph. If G is a weighted graph with positive weights on its edges (Fig. 2.16a), the spanning tree may be unweighted (if its edges are not weighted) (Fig. 2.16b) or weighted if its edges maintains the same weights they have in G. The cost of a *weighted spanning tree* is simply the sum of the weights of the $n - 1$ edges belonging to it (Fig. 2.16c, d). It is easy to see that for a given graph we may generate many spanning trees and that the cost of two weighted spanning trees may be the same, even if the trees are different. Among all the possible weighted spanning trees of a graph G, there will be some whose cost is minimum in comparison to all the others: the problem we are going to analyze is that of how to find one of these minimum-cost trees (Fig. 2.16d).

This typical graph theory problem was first confronted in 1926 by the mathematician Otakar Boruvka, during a period when, as he himself declared, "graph theory was decidedly dull". Presented with the task of minimizing the costs of constructing a grid to supply electricity to southern Moldavia, Boruvka decided

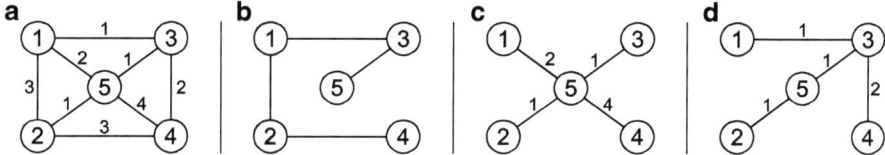

Fig. 2.16 (**a**) A connected, non-orientated graph G=(V, E) with positive weights on its edges; (**b**) a possible non-weighed spanning tree of G; (**c**) a weighted spanning tree of G whose cost is 8; (**d**) a minimum weighted spanning tree of G whose cost is 5

to tackle the problem using graphs: starting from the assumption that the junctions of the grid could only be located in cities, the nodes of the graph represented those cities to be connected to the electricity grid, while the weighted edges represented the lengths of power lines with their respective associated costs. The solution he came up with led to the definition of the minimum spanning tree, which stimulated immediate interest among the mathematics community.

Subsequently, the refinements made to the technique and its application to a wide range of problems within the field of computer science (computer design, network security, automatic language recognition, image processing, etc.) have led to the minimum spanning tree being attributed to the algorithms of Kruskal (1956) and Prim (1957) rather than to Boruvka's work (in spite of the fact that both Prim and Kruskal cited Boruvka in their papers). In the following we present Prim's greedy solution, illustrating its functioning with a specific example.

The algorithm for generating a tree starting from a graph adds just a single node and a single edge at each step. From the graph in Fig. 2.17a it can be seen that at the outset of the procedure the tree, T, that we are going to construct has just the node 1. The choice of the next new node to be added to the tree will be made by analyzing the costs of all the edges that connect the node already in E to all the other nodes remaining in the graph (nodes R in the figure). In our example, at the first step, we have only edges $(1, 2)$ and $(1, 3)$. Therefore both edge $(1, 2)$ and node 2 will be added to the tree (Fig. 2.17b). Now edges $(1, 3)$, $(2, 5)$ and $(2, 4)$ have to be analyzed; the latter two each have a cost of 1, so it makes no difference which of the two is chosen. In the example node 5 and edge $(2, 5)$ are added to the tree (Fig. 2.17c). In order to pass from the situation depicted in Fig. 2.17c to that in Fig. 2.17d it is necessary to choose the edge with the lowest cost from among the following: $(1, 3)$ of cost 2; $(2, 4)$ of cost 1; $(5, 4)$ of cost 3; and $(5, 3)$ of cost 14. The lowest cost is 1, therefore edge $(2, 4)$ and node 4 are chosen. In similar fashion we pass from the situation in Fig. 2.17d to that in Fig. 2.17e by choosing edge $(1, 3)$ and node 3, and from that in Fig. 2.17e to that in Fig. 2.17f by choosing edge $(4, 6)$ and node 6.

We repeat that this algorithm, based on the greedy technique, yields an optimal solution, and that the optimality of this solution is due to the mathematical properties of the problem. The optimal solution that is being sought by the algorithm is a spanning tree of graph G for which the sum of the weights of its edges is minimum.

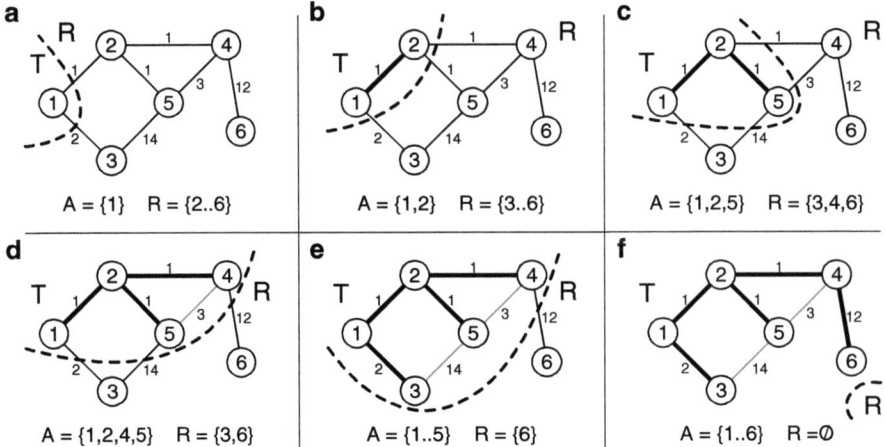

Fig. 2.17 Example of the operation of Prim's algorithm

Each greedy choice generates a minimum spanning subtree from among all the spanning subtrees covering the same portion of the graph. We now present the scheme to prove by induction that the algorithm is correct.

Basic step: A tree, T, consisting of only one node and of no edge is obviously a tree for which the sum of the weights of its edges is minimum;

Inductive hypothesis: Let T with k nodes and $k-1$ edges be a tree for which the sum of its weights is a minimum among all the possible weighted spanning trees having the same k nodes;

Inductive step: The new tree T' obtained by adding to the tree of the inductive hypothesis the edge (x, y), which is of minimum cost among all the edges with x in T and y in R, maintains the property of being of minimum cost among all the trees covering the same $k+1$ nodes.

Let us now round off this section by outlining a further application of the greedy technique, one which enables an approximate solution to a problem to be obtained when arriving at an optimum solution would require the analysis of too great a quantity of information (see Sect. 2.4).

Suppose that a house burglar, in order to facilitate his exit, can only steal items that he is able to stow away completely inside his "swag sack". Obviously, his primary aim on each break-in is to maximize the overall value of the items he is going to steal. The ideal for him would be to analyze all the possible combinations of items that could be stowed into the sack, and for each combination calculate the overall value it amounts to. However, if our burglar were to do this it would not be long before he ended up in jail!

In general terms this problem, known as the *knapsack problem*, consists of having a set of n elements (items to be stolen) each of a predetermined weight and value, and a container of capacity C (the swag sack). Given that the sum of the weights is limited by the capacity of the container, the aim is to maximize the sum of the values.[8] An exact solution can only be obtained by analyzing all the 2^n possible combinations of the given elements (using, for example, the backtracking technique), discarding those combinations whose sum exceeds the capacity C and choosing from among the remaining combinations the one which maximizes the sum of the values. However, as we will now go on to explain in the next section, the *goodness* of such an algorithm would not be acceptable. In spite of this, it is possible to prove that a greedy algorithm that analyzes the n elements in descending order of the ratio between value and weight guarantees a good approximate solution that is only slightly worse than the optimum solution. This argument will be dealt with more thoroughly in Chap. 5 (see Sect. 5.3).

2.4 How to Measure the Goodness of an Algorithm

So, what do we mean by the goodness of an algorithm? What do we mean when we say that an algorithm behaves in an efficient manner? And what do we mean when we declare that an algorithm is not acceptable?

All the examples we have seen so far have shown us that an algorithm arrives at a solution to a problem step by step, performing at each step basic operations that can be either comparative, or arithmetical, or logical. In fact, the measure of the goodness of an algorithm is calculated according to the number of basic operations it has to perform. We will now see how.

Let us start by analyzing a very simple problem: calculating the sum of the first n integer numbers, when n is given a priori. The *Sum*1 algorithm in Fig. 2.18 solves this problem correctly, following a method derived directly from the problem's definition: starting from 0 it builds to the result by summing first 1, then 2, then 3, and so on until it has added n, too.

Accepting the hypothesis that a constant time c_1 is necessary for summing any two numbers (independently of their values), we see that the time for generating the output will depend on how many numbers have to be summed (i.e., $c_1 \times n$). This might appear self-evident, given that it is intuitive to think that the time taken to perform a task is going to be the greater, the greater the number of operations that have to be performed. Nevertheless, in the following we will show that this statement is not always true. In fact, generally speaking, a number of different solutions to a problem can be derived, depending on whether the design of the algorithm is based on the simple definition of the problem, or on some intrinsic properties of the problem itself.

---

[8]This is a typical allocation problem similar to the problems of how to load crates of different weights into a container, how to cut plates of different shape from a sheet, how to store files of different sizes on a disc, etc.

> **Input**: an integer number n
> **Output**: S = the sum of the first n integer numbers
>
> **Algorithm** *Sum*1
>   **Step 1**: Set $S = 0$;
>   **Step 2**: Repeat $S = S + i$ for $i = 1,...,n$;
>   **Step 3**: Provide S in output.

Fig. 2.18 Algorithm *Sum*1

> **Input**: an integer number n
> **Output**: S = the sum of the first n integer numbers
>
> **Algorithm** *Sum*2
>   **Step 1**: $S = n \times (n+1)/2$;
>   **Step 2**: Provide S in output.

Fig. 2.19 Algorithm *Sum*2

By way of illustration of this point an anecdotal story recounts that around 1780 a primary school teacher set his pupils the task of summing the first 100 integer numbers, convinced this would keep them occupied for a few hours. Unluckily for him, however, there was a pupil, named Carl Friedrich Gauss, in the class, who, after just a few minutes, brought him the correct solution: 5050. Gauss had noticed that the sum of the first and last numbers (1 and 100), the second and penultimate numbers (2 and 99), and so on, remained constant at 101. Thus it was sufficient simply to multiply the number 101 by 50 to obtain the correct result. The algorithm *Sum*2 in Fig. 2.19 is based on the generalization of Gauss's observation:

> Whatever the value of n, in the sequence of the first n integer numbers the sum of the pair of numbers that are symmetrical with respect to $n/2$ is always $n + 1$.

The *Sum*2 algorithm correctly calculates the sum of the first n integer numbers by performing just three elementary operations (a product, a sum and a division), irrespective of the number of elements to be summed. If we call the constant times needed to calculate a product and a division, c_2 and c_3, respectively, the time required for the entire algorithm to execute will be $c_1 + c_2 + c_3$, and we say that the time complexity of the algorithm is of constant order (because it is independent of n). This contrasts with the time complexity of the *Sum*1 algorithm, which has a linear order (because its time complexity grows in proportion to the value of n). We can therefore say that *Sum*2 algorithm is more efficient than *Sum*1 in that it requires a constant time irrespective of the size of the problem, rather than a time that increases linearly with the increase in n.

Although the linear order often cannot be attained due to the time complexity of a problem, we wish to emphasize that it is possible to prove the linear time complexity

2 How to Design an Algorithm   51

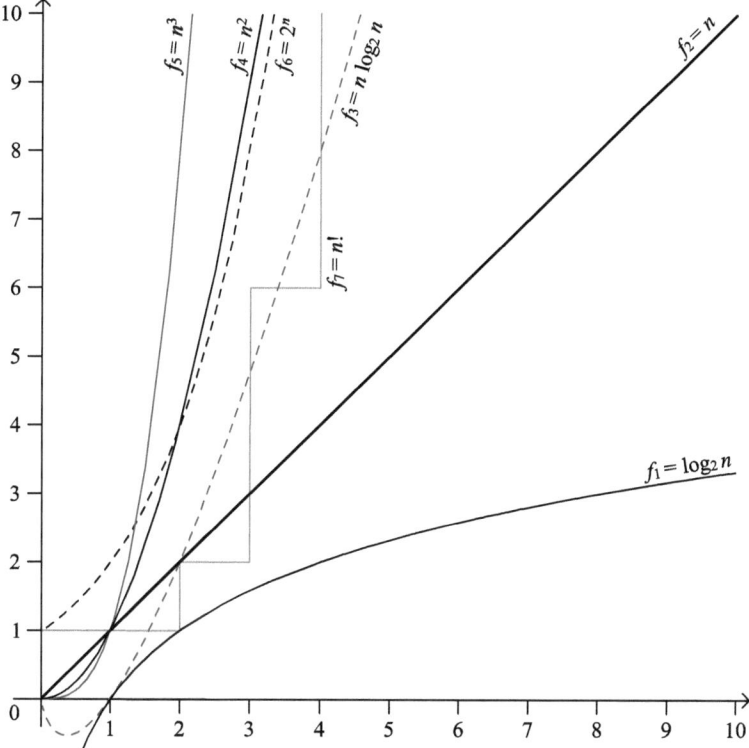

Fig. 2.20 Graphic of some important functions for calculating computational complexity

for the topological order and for the Prim algorithm. Figure 2.20 shows some of the more important functions that appear in the computation of an algorithm's time complexity. Regardless of how much the reader may remember about functions from high school studies, it should be immediately obvious to him/her that there are some functions, like *logarithm*, that increase more slowly than others as n increases, just as there are some functions that increase much, much faster (for example, f_5 and f_6 in Fig. 2.20).

For readers who may wish to enjoy playing with numbers, Table 2.1 gives the values, as n increases, of the various functions depicted in the graphic in Fig. 2.20. Here we just want to point out that executing an exponential number of operations, such as 2^{100} for instance, using today's most powerful computers would require billions of years, while a length of time equal to the age of the universe would not be sufficient to perform $n!$ operations[9] when n is quite large (just consider that 10! is already 3,628,800).

---

[9]The factorial of a number n is the product of all the whole numbers between 1 and n, inclusive. That is, $n! = n \times (n-1) \times (n-2) \times (n-3) \times \ldots \times 3 \times 2 \times 1$.

Table 2.1 Increase in values of the functions shown in Fig. 2.20. Note that compact notation has been adopted to avoid having to use an excessive number of figures to express the values, that is, $10^2 = 100$, $10^3 = 1{,}000$, $10^4 = 10{,}000 \ldots$

n	$f_1 = \log_2 n$	$f_2 = n$	$f_3 = n \log_2 n$	$f_4 = n^2$	$f_5 = n^3$	$f_6 = 2^n$	$f_7 = n!$
10	3.32	10	33.22	10^2	10^3	1,024	3.62×10^6
10^2	6.64	10^2	664.39	10^4	10^6	1.26×10^{30}	9.33×10^{157}
10^3	9.97	10^3	9,965.78	10^6	10^9	1.07×10^{301}	
10^6	19.93	10^6	1.99×10^7	10^{12}	10^{18}		
10^9	29.90	10^9	2.99×10^{10}	10^{18}	10^{27}		

Due to the foregoing considerations, problems requiring algorithms that make use of an exponential (or factorial) number of operations, such as those employing the backtrack technique, are termed *intractable*. This is because, however interesting they may be from a theoretical point of view, in practice, unless they are quite small in size, they can only be solved in an approximate manner. Therefore, it is essential that in designing an algorithm it is not just the aim of resolving the problem that is born in mind, but also, and perhaps most importantly, the efficiency of its resolution. In order to be able to measure the efficiency of an algorithm, the theory of computational complexity studies the classification of problems in terms of their intrinsic difficulty and defines as *tractable* all those problems that are amenable to resolution by at least one algorithm of polynomial time complexity. In Chap. 3, these problems will be covered in more detail, while in Chap. 10 it will be shown how introducing probabilistic concepts into algorithm design enables efficient, if not exact, solutions to be found for very difficult problems.

2.5 The Design

In this chapter, although we have covered all the fundamental aspects involved in designing an algorithm, we have not necessarily presented them in a sequence that the reader has been able to follow readily. In this final section, therefore, we will provide *a step-by-step* summary of the process, using the *binary search* procedure by way of illustration.

Let us start by summarizing the basic characteristics of an algorithm that have been presented under a variety of headings in the first two chapters of this book. Following the teachings of Donald Knuth we can say that an algorithm, when given a specific *input*, needs to generate an *output* by executing a series of successive steps (a sequence of elementary operations) that have the following properties:

Finiteness: An algorithm must always terminate after the execution of a finite number of steps. It is worth noting that sometimes, even if rarely, the non-termination of an algorithm is acceptable, though in such cases we speak of computational procedure rather than algorithm. An example of a computational procedure is a computer operating system, designed to control the execution of other programs and to remain suspended in a state of readiness when no program is being executed.

Effectiveness: Every algorithm must be effectively executable, that is, each operation must be sufficiently basic as to be executed using "pen and paper" in a finite amount of time. A typical example of lack of effectiveness would be the division of a natural number by zero.

Definiteness: Each step of an algorithm must be defined in a clear, unambiguous way, that is, it must give rise to the same sequence of operations and outcomes regardless of by whom, and when, it is being executed. As an example of lack of definiteness just think of the way culinary recipes are typically presented, with frequent use of terms such as "according to taste or as much as is needed", and which, therefore, cannot be considered algorithms.

An algorithm can thus be expressed in any language, as long as this language can guarantee definiteness. *Programming languages*, which are designed to ensure that every statement has a unique meaning when it comes to be interpreted by the computer, were introduced to avoid ambiguities that might arise from the use of natural languages. An algorithm expressed by means of an appropriate programming language is called a *program*. Once an algorithm's design is complete, its *correctness* needs to be tested, that is, it has to be proved that the algorithm provides the correct output for every possible input. It needs emphasizing, however, that this proof concerns the "philosophy of the algorithm" only, and thus it must be independent of whichever programming language has been used to implement the algorithm. The *validation* therefore concerns the algorithm, not the program. Once the algorithm has been validated it can be transformed into a program. This program must then, in its turn, also be *certificated*, that is, it has to be verified that the program correctly expresses the algorithm it is intended to implement.

To illustrate what we have just summarized, let us consider the following problem: *given a sequence A of n elements, each distinct from the other, and a known value x, check whether x is in the sequence A and provide as output a value j which indicates the position of x in A, if x is in the sequence, or output 0 if it is not*. For example, given the sequence $A = (2, 8, 1, 7, 5)$ when $x = 8$, the output will tell us that this value of x is present in the sequence and is located in the second position ($j = 2$). Instead, when $x = 4$, the output will be $j = 0$, indicating that the element is not present in the sequence.

Let us now see how it is possible to solve the problem without having any further information about the input: we have to compare the element x with the first element of the sequence, then with the second, and so on until either x has been found, or all of the elements in A have been analyzed. The method of sequential search presented here is easy to validate, given that an exhaustive search on all the elements in the sequence makes it possible to say that the element x has not been found. We can therefore pass from the algorithm to the program, which is shown in Fig. 2.21.

A vector data structure has been chosen for the set A because this offers the simplest way of accessing the elements of the sequence by indices. The repetition of step 2 until either the searched-for element has been found, or one or more elements

> **Input:** A: vector of n elements; x: a known element
> **Output:** j if $x = A(j)$ $(1 \leq j \leq n)$; $j = 0$, otherwise
>
> **Algorithm Linear Search**
>   **Step 1:** Set $j = 1$
>   **Step 2:** Repeat $j = j + 1$ until $j > n$ or $x = A(j)$
>   **Step 3:** If $j > n$ then set $j = 0$
>   **Step 4:** Provide j as output

Fig. 2.21 Linear search program

remain to be analyzed ($j < n$), guarantees that the program has translated the algorithm in a correct manner. Note also that the algorithm (and thus the program) is based solely on elementary operations that are clearly defined and executable in practice. And at this point we can say that the algorithm is finite, given that its computational complexity is of order n, that is, the number of elements to be compared increases as the size of the sequence increases: for these reasons this type of algorithm is also known as a *linear search algorithm*.

In order to optimize the design of an algorithm it is essential to exploit to the full all information concerning its input. To this end, let us see how the search algorithm might be modified if we accept the hypothesis that the set of information in input is sorted. Thus, instead of performing comparisons starting with the first element and then proceeding by comparing x with all the elements of the sequence, the fact that the input is sorted enables us to start by comparing x with the element in the middle of the sequence, say $A(middle)$. Then, even if the result of this comparison is shown that j is not the index of the middle element, this single comparison enables us to halve the number of elements that need to be analyzed. In fact, $x < A(middle)$ means that x, if it exists in the sequence, can only lie to the left of $A(middle)$. Similarly, if x is greater than $A(middle)$, then, it can only lie to the right of $A(middle)$. This reasoning is now applied to the half of the sequence just identified and a *new middle* is computed. The algorithm then repeats this step until the position of element x in the sequence is found, or until there are no more elements to be compared. Figure 2.22 shows the complete example of the search for the element 20 in the sequence $A = (-2, -1, 3, 5, 7, 9, 15, 20, 22, 25, 33, 35)$.

The algorithm in Fig. 2.22 is known as the *binary search algorithm*. It can be validated through a *a proof by contradiction*, that is, by demonstrating that declaring the algorithm does not function correctly calls into question the conditions imposed by the formulation of the problem (see Fig. 2.23).

Figure 2.24 shows the transformation of the binary search algorithm into a program (in terms of our simplified programming language).

The program is effectively concentrated in Step 2, which correctly transforms the binary search algorithm because it continues to check the middle element of the current subvector until either it finds the element in the sequence ($j \neq 0$), or there are no more elements to check, that is, the end points of the sub-vector

> **Step 1:** $A(1,\ldots,12) = (-2,-1,3,5,7,9,15,20,22,25,33,35)$ and $x = 20$ is compared against the middle element, $A(6) = 9$.
> 20 is greater than 9, therefore in the next step x will be searched for to the right of 9.
> **Step 2:** $A(7,\ldots,12) = (15,20,22,25,33,35)$ and $x = 20$ is compared against the new middle element $A(9) = 22$.
> 20 is less than 22, therefore in the next step x will be searched for to the left of 22.
> **Step 3:** $A(7,8) = (15,20)$ and $x = 20$ is compared against the new middle element $A(7) = 15$.
> 20 is greater than 15, therefore in the next step x will be searched for to the right of 15.
> **Step 4:** $A(8) = (20)$ and $x = 20$ is compared against the new middle element $A(8) = 20$.
> The two numbers are equal, thus x lies in the eighth position in the sequence and the output is $j = 8$.

Fig. 2.22 Binary search algorithm

> Rejecting the correctness of the algorithm is equivalent to stating that:
>
> 1. x is greater than *A(middle)*;
> 2. x is in the sequence;
> 3. x is to the left of *A(middle)*.
>
> Statements 1, 2 and 3 can only be simultaneously true if values that are equal exist in the sequence or it is established that the values in the sequence are not sorted. In either of these cases a part of the information provided as input is contradicted and is therefore in absurdum.

Fig. 2.23 Validation of the binary search algorithm through proof by contradiction

under analysis switch places ($lx > rx$). The detection of the subvector is performed correctly because it is in agreement with the validation of the algorithm (Fig. 2.23).

All that now remains to complete this outline of the binary search algorithm is to calculate its *computational complexity*. To do this, let us express the problem in terms of trees. The elements of a sorted sequence can be represented on a binary tree in a recursive manner bearing in mind that:

- The root of the tree contains the middle element of the sequence;
- The left subtree represents the subsequence to the left of the middle element;
- The right subtree represents the subsequence to the right of the middle element.

The sorted sequence of Fig. 2.22 is represented on the binary tree in Fig. 2.25. Highlighted on the tree are those nodes of the sequence that the algorithm has compared with the element $x = 20$. From analysis of this tree it can be seen that each individual step of the binary search algorithm on the sorted sequence corresponds to individual choices on a decision tree representing this sorted sequence. The maximum number of comparisons executed by the algorithm is therefore equal to the height of the tree (one for each node on the path from the root to a leaf), plus 1 extra comparison in the case that element x has not been found in the sequence, whereupon it is necessary to check that there are no further

Input: A : a vector of n distinct elements sorted in non-decreasing order; x: a known element
Output: j if $x = A(j)$ ($1 \leq j \leq n$,); $j = 0$, otherwise

Algorithm Binary Search
  Step 1: Set $j = 0$, $lx = 1$, $rx = n$
  Step 2:
    Repeat
      $middle = (lx + rx)/2$
      if $x = A(middle)$ then $j = middle$ otherwise
        if $x < A(middle)$ then $rx = middle - 1$ otherwise
          $lx = middle + 1$
    until $lx \leq rx$ and $j = 0$
  Step 3: j is provided in output

Fig. 2.24 Binary search program

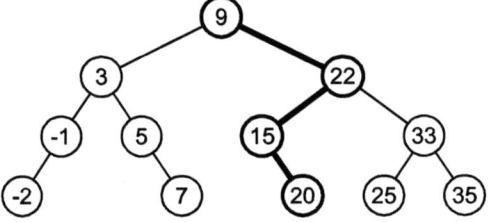

Fig. 2.25 Decision tree for the search for element 20 in the sequence A= $(-2, -1, 3, 5, 7, 9, 15, 20, 22, 25, 33, 35)$

elements to be compared. Since the height is equal to $\log_2 n$, it follows that the maximum number of comparisons performed by the binary search algorithm on a sorted sequence of n elements will be at most ($\log_2 n + 1$), and we say that the computational complexity of the algorithm is *logarithmic*.

The binary search algorithm is a typical example of an algorithm based on the *divide and conquer* technique, which, following the well-known ancient Roman battle strategy, first divides the instance of the given problem into subinstances (usually two, and of equal dimensions if possible), and recursively solves each instance separately. At the end of the process the solutions of the subinstances are combined, if necessary, in order to generate a solution to the original problem. As we showed in the case of the binary search, this approach, when applicable, leads to efficient solutions, given that the division into subinstances of equal dimensions usually enables the concept of logarithm to be introduced into the calculation of the computational complexity.

2.6 Bibliographic Notes

A wide range of introductions to algorithm design are to be found in the literature.

Knuth's book [68] has been amply cited in the course of this chapter. For indicative purposes only, and without wishing to imply it makes the list any way exhaustive, we mention two further books that have become classics: Aho et al. [3], concentrating on design aspects, and [62] focusing on data structures. Among the more recent books, again purely by way of indication, we point out [67] and [71]. Finally, we would like to recommend Harel's rather more popularized book [55], which is now in its third edition and has been translated into many languages, including Chinese.

The form of the eight queens problem presented in this chapter takes its inspiration from the treatment of Wirth in [111]. An enjoyable account of the developmental history of the solution to the problem of the minimum spanning tree is to be found in [52], while we point readers interested in the matching problem to [102].

For information on the life and works of Euler we indicate the website, *The Euler Archive*, which also provides access to his original publications: www.math.dartmouth.edu/~euler/.

Pólya, in his seminal book on inductive reasoning in mathematics [92], provides a detailed description of mathematical induction, devoting an entire chapter to the subject.

Finally, as regards codes, the reader is pointed to [107].

Chapter 3
The One Million Dollars Problem

Alessandro Panconesi

Dedicated to the memory of a gentle giant

Abstract In May 1997, Deep Blue, an IBM computer, defeated chess world champion Garry Kimovich Kasparov. In the decisive Game 6 of the series, Kasparov, one of the most talented chess players of all time, was crushed. According to Wikipedia "after the loss Kasparov said that he sometimes saw deep intelligence and creativity in the machine's moves". Can machines think? Scientists and philosophers have been debating for centuries. The question acquired powerful momentum in the beginning of the twentieth century, when the great mathematician David Hilbert launched an ambitious research program whose aim was to prove conclusively that his fellow mathematicians could be supplanted by what we now call computers. He believed that mathematics, one of the most creative human endeavours, could be done by them. In spite of the rather dramatic denouement of Hilbert's program, the question he posed is not yet settled. The ultimate answer seems to be hidden in a seemingly silly algorithmic puzzle: a traveling salesman is to visit N cities; can the cheapest tour, the one that minimizes the total distance travelled, be computed efficiently?

Our story begins in Paris on March 24, 2000, when the Clay Foundation, during a ceremony held at the Collège de France, announced an initiative intended to *"celebrate the mathematics of the new millennium."* The American foundation instituted seven prizes of one million dollars each for the resolution of seven open problems of mathematics, selected by a committee of distinguished mathematicians for their special significance. Among these seven "Millennium Problems", alongside

---

A. Panconesi (✉)
Dipartimento di Informatica, Sapienza Università di Roma, via Salaria 113, 00198 Roma, Italy
e-mail: ale@di.uniroma1.it

classics such as the Riemann Hypothesis and the Poincaré Conjecture, a newcomer stood out, a computer science problem known as *"P vs. NP"*. Formulated just 40 years ago in the Soviet Union by Leonid Levin, a student of the great Andrei Kolmogorov, and simultaneously and independently in the West by Stephen A. Cook of the University of Toronto in Canada, and by Richard M. Karp of the University of California at Berkeley, this algorithmic problem is considered to be of fundamental importance not only for the development of mathematics and computer science but also for its philosophical implications. Its resolution will tell us something fundamental about the intrinsic limitations of computers, regardless of the state of technology. At first blush, the problem does not appear forbidding or especially noteworthy but, as they say, looks can be deceiving. One of the many ways in which it can be formulated is the following, known as the Traveling Salesman Problem (henceforth abbreviated as TSP): suppose that we have a map with N cities that must be visited by means of the shortest possible route. That is, starting from any city, we must visit each city once and only once, and return to the city we started from. Such an itinerary is called a tour and the goal is to find the shortest possible one.

Computers seem ideally suited for such tasks. The resulting mathematical challenge is to find an algorithm (or, in the final analysis, a computer program) that is able, given any input map, to find the shortest tour. The crucial requirement that our algorithm must satisfy is to be efficient. The one million dollars problem is the following: does there exist such an algorithm?

Although it may seem surprising, no one has ever managed to find an efficient algorithm for the TSP, and it is conjectured that none exists! Note that it is completely trivial to develop an algorithm that finds the shortest tour. Already in the first year of college computer science students are typically able to come up with this algorithm:

> Enumerate all possible tours among the N cities, measuring the length of each one of them, and record the shortest one observed so far.

In other words, if we generate by computer all possible tours, measure the length of each, and keep the shortest distance observed so far, we will eventually find the shortest one. The catch, of course, is that this algorithm, as easily programmable as it may be, is not efficient: its execution time is so high to make it completely useless. To see this, note that if you have N cities the number of all possible tours is given by:

$$(N-1) \times (N-2) \ldots \times 4 \times 3 \times 2 \times 1.$$

(Start from any city, then we have $N-1$ choices for the second city, $N-2$ for the third, and so on). This number is denoted as $(N-1)!$ and, even for small values of N, it is so large as to require a certain amount of thought in order to be comprehended. For example, 52! is equal to:

80,658,175,170,943,878,571,660,636,856,403,766,975,289,505,440,883,277,824,000, 000,000,000[1]

How big is this number? Expressed in billionths of a second it is at least 5,000 billion times greater than the age of the universe. In industrial applications it is necessary to solve the traveling salesman problem for values of N in the order of the thousands, and beyond (Fig. 3.1). Numbers like 5,000! or 33,474! would require quite some time and ingenious mathematical notation to be grasped and are immeasurably greater than a puny, so to speak, 52!. It is therefore quite clear that the algorithm described is of no help in solving realistic instances of the TSP. Moreover, and this is a fundamental point, *no technological improvement will ever make this algorithm practical.*

In summary, the difficulty with the TSP is not that of finding an algorithm that computes the shortest tour, but to find an efficient one. Surprisingly, nobody has been able to find such a method and not for lack of good-will. The problem is quite relevant from the point of view of the applications, where it is encountered often and in different guises. For decades, researchers have tried to come up with efficient algorithms for it, but without success.

In conclusion, the algorithmic problem for which one million dollars are at stake is this: find an efficient algorithm (i.e., a programmable procedure) for the TSP, or prove that such an algorithm does not exist. This deceptive-looking problem is at the heart of a series of far-reaching mathematical, technological and even philosophical issues. To understand why we return to Paris, but going back in time.

3.1 Paris, August 8, 1900

> Who of us would not be glad to lift the veil behind which the future lies hidden; to cast a glance at the next advances of our science and at the secrets of its development during future centuries? What particular goals will there be toward which the leading mathematical spirits of coming generations will strive? What new methods and new facts in the wide and rich field of mathematical thought will the new centuries disclose?

Thus began the famous keynote address that the great mathematician David Hilbert, at age 40, gave at the International Congress of Mathematicians in 1900.[2] With his speech Hilbert who, along with Henri Poincaré, was considered the most influential mathematician of the time, outlined an ambitious program to shape the future of mathematics. The centerpiece of his proposal was his now famous list of 23 open problems that he considered to be central to the development of the discipline.

---

[1] The calculation, by no means trivial, is taken from http://www.schuhmacher.at/weblog/52cards.html
[2] Maby Winton Newson [85].

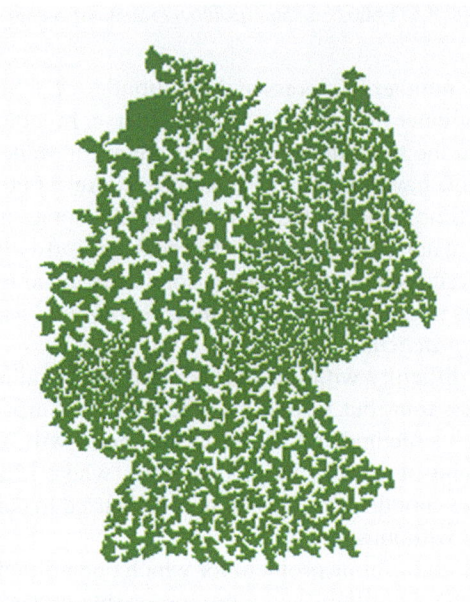

Fig. 3.1 A TSP input instance consisting of 15,112 German cities. The calculation of the shortest tour (shown in the figure) kept two teams of researchers from Rice University and Princeton University busy for several months and required considerable ingenuity and computing resources: 110 parallel processors, equivalent to 22.6 years of calculation on a single processor. In 2006, another instance with 85,900 cities was solved, again with massive parallel resources. The computation time, translated for a single processor, was 136 years. (See "Traveling salesman" on Wikipedia). Note that this considerable deployment of resources was needed to solve just two isolated instances of the problem – the methods developed have no general applicability

What makes this speech so memorable in the history of ideas is the fact that the twentieth century has been a golden age for mathematics. With his 23 problems Hilbert succeeded in the incredible undertaking of shaping its development to a large extent for a 100 years. On the problems posed by Hilbert worked titans such as Kurt Gödel, Andrei Kolmogorov and John (Janos) von Neumann, and more than once the Fields Medal, the highest honor in mathematics, has been assigned for the resolution of some of Hilbert's problems. In fact, the event that we mentioned at the beginning, the announcement of the seven Millennium Problems at the Collège de France, was meant to celebrate and hopefully repeat the fortunate outcome of that memorable speech, exactly 100 years later.

In modern terms, one of Hilbert's main preoccupations, which was already apparent in his work on the foundations of geometry *"Grundlagen der Geometrie"* (Foundations of Geometry) and that became more explicit with the subsequent formulation of his program for the foundations of mathematics, was the pursuit of the following question: can computers do mathematics, and if so to what extent?

One of the underlying themes of Hilbert's vision was an ambitious program of "mechanization" or, in modern terms, computerization of the axiomatic method, one of the pillars of modern science that we have inherited from Greek antiquity and of which "The Elements" by Euclid, a book that remained a paragon of mathematical virtue for many centuries, is probably the most influential example. The axiomatic method, in no small measure thanks to the revitalizing work of Hilbert, now permeates modern mathematics. According to it, mathematics must proceed from simple assumptions, called axioms, by deriving logical consequences of these through the mechanical application of rules of inference. To give an idea of what it is about let us consider a toy axiomatic system whose "axioms" are the numbers 3, 7 and 8 and whose "inference rules" are the arithmetic operations of addition and subtraction. Starting from the axioms, we apply the inference rules in every possible way to generate other numbers. For instance, we can derive $5 = 8 - 3$, and $2 = 8 - (3 + 3)$, etc. The new numbers thus obtained are the "theorems" of this theory. The "theorems" obtained can be combined with each other and with the axioms to generate other "theorems" and so on. In our example, since 2 is a theorem and + is an inference rule, we can generate all even numbers. Thus 4, 6, etc., are theorems whose corresponding proofs are $2 + 2$, $2 + 4$, and so on. The axiomatic systems that are used in mathematics are obviously much more expressive and complex and, in general, are meant to manipulate predicates – logical formulas that can be true or false – rather than raw numbers. Predicates are sentences that express the universe of mathematical statements, things like *"There are infinitely many prime numbers"*, *"The equation $x^n + y^n = z^n$ has no integer solutions for $n > 2$"*, and so on. But, by and large, the essence of the axiomatic method is the one described: precise inference rules are applied to predicates in a completely circumscribed and mechanical fashion, in the same way as the four arithmetic operations handle numbers. The axiomatic method offers the best possible guarantees in terms of the correctness of the deductions. If the starting axioms are propositions known to be true, and they do not contradict each other, by applying the rules of inference (which are nothing other than simple rules of logical deduction) we can only generate other true statements. For the Greeks the axioms were self-evidently true by virtue of their sheer simplicity. (In contrast, the modern point of view requires only that they do not contradict each other.) For instance, one of the axioms of Euclidean geometry is the following, arguably self-evident, statement: *"If we take two points in the plane, there is a unique straight line that goes through them"*.[3] The mechanical aspect of the axiomatic approach is clear: given that the application of an inference rule is akin to performing an arithmetical operation, and since a mathematical proof is nothing other than the repeated application of such elementary and mechanical operations, that from a set of premises lead to a conclusion, a computer should be able to reproduce such a process. And indeed, a computer is certainly able to mimic it. For instance, given

---

[3] Of course, as we now know with hindsight 2,000 years later, assuming that axioms are self-evident truths is a slippery slope, exemplified by the development of non-Euclidean geometry.

an alleged proof of a mathematical statement such as *"There exist infinitely many prime numbers"* a computer can rather easily determine if the given proof is correct or not. But beware! This by itself does not mean that computers can supplant mathematicians. The aspects of the matter are quite complex, but for our purposes it is sufficient to focus on a fundamental dichotomy, the difference between inventing and verifying. To check the correctness of a given mathematical proof – i.e. its verification – is a mental activity fundamentally quite different from the act of finding it. The latter is a creative act while the former is not.

To create a mathematical proof is a convoluted and complex psychological process in which logic does not play a primary role at all, necessarily supplanted by mysterious visions that suddenly emerge from the subconscious. In contrast, to verify a demonstration (i.e. to convince oneself of its correctness) is a task that can be intellectually very demanding but that does not require creativity. In order to verify a proof one must check that the conclusions follow, now logically, from the premises through the correct application of rules of inference. This is an essentially mechanical process, somewhat similar to performing a long series of complicated arithmetical operations. It is in this second phase, during which the correctness of the visions and insights of the creative process of mathematical discovery are put under rigorous scrutiny, that the task of logic, very limited in the discovery phase, becomes predominant. In the words of Alain Connes, the great French mathematician, winner of the Fields Medal,

> Discovery in mathematics takes place in two phases. [...] In the first, the intuition is not yet translated into terms that you can communicate in a rational manner: here is the vision, [...] a kind of poetic inspiration, which is almost impossible to put into words. [...] The second phase, that of demonstration, is an act of verification: it requires high concentration and a kind of extreme rationalism. But luckily there is still a vision, which activates the intuition of the first phase, does not obey certainty and is more akin to a burst of poetic nature.[4]

By transcribing the complex creative process that leads to the discovery of mathematical truths in a succession of elementary and verifiable steps, the mathematician submits voluntarily to the most exacting verification discipline: the tortuous path that leads to the inescapable conclusion is laid out carefully and thoroughly in a long sequence of logical, and hence unimaginative and ultimately mechanical, steps. In such a way not only a machine, but even a person who is completely devoid of mathematical talent can be convinced!

Unfortunately, the teaching of mathematics, even at university level, is almost exclusively confined to the hard discipline of verification. This hides its creative dimension and it is not surprising that mathematics suffers from the undeserved reputation of being a dry discipline. That one of the most imaginative human activities must suffer from such an injustice is a tragedy comparable to the destruction of the great masterpieces of art.

---

[4] Alain Connes [19].

3.2 "Calculemus!"

Let us go back to Hilbert and his influential school, who believed that it was possible to push the level of formalization of the axiomatic method to a point where it becomes entirely mechanical. It may seem surprising now, but Hilbert and his followers, including geniuses like von Neumann, not only believed that the verification process but also the discovery of mathematical truths could be made by a computer! As we discussed, to find a mathematical proof and to verify it are two completely different psychological processes. How could Hilbert think that even the former, a quintessentially creative act, could be carried out by a machine? In Science, often the power of an idea stems not so much from the fact that it ultimately turns out to be correct, but rather from its vision, its power to summon *"the leading [minds] of coming generations"* who, in their quest for truth, will produce revolutionary new knowledge, and a leap forward in the understanding of the world. Hilbert's Program for the "computerization" of mathematics is a case in point. Its great merit was to frame a set of bold, profound questions of great mathematical, and even philosophical, import in such a way that they could be fruitfully tackled in precise and concrete mathematical terms. Hilbert's promethean conjectures gave impetus to a steady flow of sensational results on the foundations of mathematics that culminated, in 1931, with the Incompleteness Theorem of Kurt Gödel, one of those achievements that, like the great masterpieces of art and music, will forever continue to inspire humankind to accomplish great and noble deeds: *"Kurt Gödel's achievement in modern logic is singular and monumental – indeed it is more than a monument, it is a landmark which will remain visible from afar, in space and time."* (John von Neumann).

One of the ways in which Hilbert's program concretely took shape is the so-called "problem of decidability". Leaving aside several important technical issues, in a nutshell Hilbert believed that one could design an algorithm (or, equivalently, write a computer program) which, on receiving as input a mathematical statement such as *"There are infinitely many prime numbers"* or *"There exist planar graphs that are not 4-colorable"*, would correctly output the correct answer – in our examples, respectively, *"Yes, this sentence is true"* and *"No, this statement is false"*. The question of whether such an algorithm exists is called the *"Entscheidungsproblem"*, the Problem of Decision. In fact, Hilbert did not consider it quite as an open problem but almost as a foregone conclusion. Hilbert believed that computers could not only verify the correctness of mathematical proofs, but find them! This is a less ambitious version, restricted to mathematics "only" of the dream cherished by the mathematician, philosopher, naturalist, diplomat and... computer scientist Gottfried Wilhelm Leibniz who thought that it was possible to settle differences of opinion through logic, by means of a computer:

> The only way to rectify our reasonings is to make them as tangible as those of the Mathematicians, so that we can find our error at a glance, and when there are disputes

among persons, we can simply say: Let us calculate [calculemus], without further ado, to see who is right.[5]

In order to address the Entscheidungsproblem it was first of all necessary to understand what an algorithm is. In the early decades of the twentieth century, several people, including Emil Post, Alonzo Church, and Kurt Gödel captured the intuitive notion of algorithm with precise mathematical definitions. But it was Alan Turing, perhaps the first genuine computer scientist, who proposed the most satisfactory characterization. Turing devised an abstract machine, now known as the Turing Machine that, to all intents and purposes, from the conceptual point of view is a full-fledged, modern digital computer capable of executing any computer program.[6] In modern terms, for Turing an algorithm was nothing other than a computer program written in a programming language like C, Java or Python. In other words, an algorithm is any process of calculation that is programmable on a modern computer. This was a rather exceptional contribution for the simple reason that at the time computers did not exist. To develop his theory Turing invented an abstract digital computer along with a primitive but powerful programming language, a sort of assembly language with which one can express, or more precisely program, any computation. The beauty and power of the approach were such that it allowed Turing to demonstrate by means of an uncannily simple argument that the Entscheidungsproblem has a flatly negative answer: there is no algorithm that can determine whether mathematical statements given in input are true or false.[7] Now, someone might breathe a sigh of relief – human creativity is saved! But not quite: as we shall see, the computer is a tough nut to crack.

Although ultimately the research program advocated by Hilbert proved to be fundamentally flawed, to the point that, essentially, the truth turned out to be the exact opposite of what he believed, the amazing discoveries made in its wake opened grand (and in some sense humbling) new vistas and jump-started whole new disciplines. In particular, although the advent of the digital computer cannot be attributed solely to the research related to Hilbert's program, there is no doubt that it played a key role. The ENIAC, perhaps the most successful early electronic computer,[8] was developed at Princeton in the context of a project supervised by

---

[5]Cited by Wikipedia's article on Leibniz: http://en.wikipedia.org/wiki/Gottfried_Wilhelm_Leibniz

[6]Turing invented two abstract devices, the Turing machine, which corresponds to a computer running a specific program, and the universal Turing machine, corresponding to a general purpose computer able to execute any program. See Chap. 1.

[7]The same conclusion was reached independently by Alonzo Church at Princeton about a year earlier thanks to the invention of the λ-calculus, that in essence is a functional programming language of which LISP is perhaps the best-known example.

[8]The first electronic computer is due to Konrad Zuse, who set one up in Berlin in 1941. In a remarkable instance of scientific short-sightedness his invention was considered by the German authorities to be "strategically irrelevant" and not developed. The indomitable Zuse immediately after the war founded a company, the second ever to market the computer, trying to plant a fertile seed in the arid soil of post-war Europe. Meanwhile, overseas the first computers were triggering the computer revolution.

John von Neumann who, besides being an early believer of Hilbert's program, knew well all developments arising from it, including those due to Alan Turing. The two became acquainted in Princeton where Turing spent some time as a post-doctoral researcher. Moreover, the difference between a Turing Machine and a modern computer is "only" technological, and indeed immediately after the end of the war Turing was able to develop in Britain a modern electronic computer called ACE.

3.3 Finding Is Hard: Checking Is Easy

The dichotomy concerning mathematical proofs that we introduced, the mechanical nature of the verification process versus the complex creativity of mathematical discovery, is at the core of the one million dollars problem. To understand why, let us consider the following algorithmic problem known as Satisfiability (henceforth SAT). We shall adopt the convention normally used in college textbooks, where an algorithmic problem is described by specifying the input data and the type of result you want in output:

> **Satisfiability (SAT)**
> **Input**: a Boolean formula $F(x_1, x_2, \ldots, x_N)$ with N variables $x_1, x_2, \ldots, x_N$, each of which may assume either the value true or the value false.
> **Question**: is the formula F satisfiable, i.e., is there a truth assignment to the variables that makes F true?

SAT is the propositional-logic counterpart of the familiar algebraic problem of finding the zeroes of a polynomial. Given a polynomial like $x^2 - 2xy + y^2 = 0$ or $x^2 + 3y(1 - z) + z^3 - 8 = 0$ one seeks values for the variables x, y, and z that verify the expressions. With SAT, logical connectives (and, or, implies, etc.) are used instead of arithmetic operations and the variables can assume the values true or false instead of numerical values. An assignment is sought that makes the formula true (see box "Satisfiability of Boolean expressions").

> **Satisfiability of Boolean expressions**
> This computational problem concerns so-called Boolean expressions, an example of which is the following:
>
> $$((x_1 \vee x_3 \to \neg x_2) \to (x_2 \to x_5 \wedge x_4)) \wedge \neg(x_1 \vee (\neg x_5 \wedge x_3)).$$

(continued)

(continued)

The variables x_1, x_2, x_3, x_4, x_5 may assume either the value T (true) or F (false). The symbols $\to, \neg, \wedge, \vee$ are logical (also called Boolean) *connectives* whose semantics is defined in the following truth tables. These are just a few of all logical connectives. The negation inverts the truth value of the expression it operates on (an expression can be a single variable or a Boolean formula, e.g. $(\neg x_5 \wedge x_3)$, $(x_2 \to x_5 \wedge x_4)$, etc.):

E	$\neg E$
T	F
F	T

The implication forbids us to derive falsehoods from truth (A and B denote Boolean expressions):

A	B	$A \to B$
T	T	T
T	F	F
F	T	T
F	F	T

And finally the disjunction (or) is true when at least one of the *operands* is true, while the conjunction (and) is true only when both operands are true:

A	B	$A \vee B$
T	T	T
T	F	T
F	T	T
F	F	F

A	B	$A \wedge B$
T	T	T
T	F	F
F	T	F
F	F	F

The above expression is satisfiable with the truth assignment $x_1 = x_2 = x_3 = F$ and $x_4 = x_5 = T$. An example of unsatisfiable formula is the following:

$$(x_1 \vee x_2) \wedge (\neg x_1 \vee x_2) \wedge (x_1 \vee \neg x_2) \wedge (\neg x_1 \vee \neg x_2)$$

(the reader should verify this).

The natural algorithmic problem associated with SAT is the following: find an *efficient* algorithm such that, given a logical formula as input, it determines whether the formula is satisfiable or not.

SAT is an important computational problem for several reasons. From a practical standpoint, many application problems can be expressed in a natural way as a satisfiability problem. Making sure that the complex software systems that control an Airbus or the railroad switches of a busy railway station do not contain dangerous programming bugs boils down to a satisfiability problem. More generally, model checking, the discipline whose aim is to ascertain whether a specific software system is free from design errors, is, to a large extent, the quest for algorithms as efficient as possible for SAT. An efficient algorithm for SAT therefore would have a significant industrial impact. The state of the art with respect to the algorithmics of SAT is rather fragmented. There are computer programs, so-called SAT-solvers, that work quite well for special classes of formulas. But apart from such very restricted situations, all known algorithms run in exponential time and hence perform very poorly in general.

SAT encapsulates well the dichotomy between finding and checking. Given a truth assignment, it is child's play for a computer to determine whether it satisfies the formula. In order to check whether a polynomial is zero, one just replaces the numerical values specified for the variables and performs simple algebraic operations. Similarly, with a Boolean formula one just replaces the truth values specified for the variables and performs a series of logical operations to verify if the formula is satisfied. Computers can do this very efficiently, in time proportional to the length of the formula. The verification of a given truth assignment, therefore, is a computationally easy task. An entirely different matter is to find a truth assignment satisfying the given input formula, if it exists. Note that, similarly to what we saw for TSP, in principle such a finding algorithm exists. Since each variable can take only the two values true or false, a formula with N variables has 2^N possible assignments. It is easy to write down a computer program that generates all these assignments and checks them one by one. If one of them satisfies the formula we have found a satisfying assignment, otherwise, if none does, we can conclude that it is unsatisfiable. But here too the same caveat that we discussed for TSP applies: 2^N is too huge a number, even for small values of N (see box "Exponential growth is out of reach for technology"). The enumerative method proposed is thus completely useless, and it will remain so forever since no improvement in computer technology can withstand exponential growth. To appreciate how fast the function 2^N grows the reader can try to solve the following fun exercise: if you fold a sheet of paper 50 times, thereby obtaining a "stack of paper" 2^{50} times thicker than the original sheet, how high would it be? The answer is really surprising!

Exponential Growth Is out of Reach for Technology

Imagine we have designed three different algorithms for a computational problem whose complexities (running times) grow like N, N^2 and 2^N, where N is the input size in bits. The next table compares their running times when the algorithms are run on a processor capable of performing 200 million operations per second, somewhat beyond the current state of the art. The acronym AoU stands for "Age of the Universe":

(continued)

(continued)

	$N = 10$	$N = 100$	$N = 1,000$
N	0.056 ns	0.56 ns	0.0056 µs
N^2	0.056 ns	0.056 µs	0.0056 ms
2^N	0.057 µs	> 16 AoU	> 1.39 10^{272} AoU

It is instructive to see what happens to these running times when the technology improves. Even with computers one billion times faster – quite a significant leap in technology! – the exponential algorithm would still take more than 10^{261} times the age of the universe for $N = 1,000$. Keep in mind that in realistic applications N is of the order of millions if not more. Technology will never be able to match exponential growth.

Thus satisfiability embodies the following dichotomy: while it is computationally easy to check if a given truth assignment satisfies the input formula, no one knows whether it is possible to efficiently find such an assignment. The dichotomy is reminiscent of a familiar situation for students of mathematics and mathematicians. To check whether a given proof is correct (e.g., to study the proof of a theorem) is a task that may be quite demanding, but that is relatively easy if compared with the conceptual effort needed to find, to create, a mathematical proof. This similarity with the computational tasks of finding and verifying may seem shallow and metaphorical but, in fact, it is surprisingly apt. As we shall see, the problem of devising an efficient algorithm for satisfiability is equivalent to that of finding a general procedure capable of determining whether any given mathematical statement admits a proof that is understandable by the human mind!

Another surprising fact is that SAT and TSP from a computational point of view are equivalent, different abstractions that capture the same phenomenon. There is a theorem, whose proof is beyond the scope of this book, which shows that if there were an efficient algorithm for SAT, then there would exist one for TSP too, and vice versa. Therefore, an equivalent formulation of the one million dollars problem is: find an efficient algorithm for SAT or prove that such an algorithm does not exist.

3.4 The Class NP

The dichotomy "checking is easy" vs. "finding is difficult" is shared by a large class of computational problems besides SAT. Let us see another computational problem that exhibits it:

> **Partitioning**
> **Input**: a sequence of N numbers $a_1, a_2, \ldots, a_N$
> **Question**: can the sequence be partitioned, i.e. divided into two groups in such a way that the sum of the values in one group is equal to the sum of the values in the other group?

For example, the sequence 1, 2, 3, 4, 5, 6 is not partitionable (because the total sum of the numbers in the whole sequence is 21, which is not divisible by 2), while the sequence 2, 4, 5, 7, 8 can be partitioned into the two groups 2, 4, 7 and 5, 8. Indeed, $2 + 4 + 7 = 5 + 8$. PARTITIONING is a basic version of a fun problem in logistics called KNAPSACK: we are given a knapsack (which can be thought of as an abstraction for a truck, a ship or an aircraft) of maximum capacity C, and a set of N objects, each of which has a weight and a value. The goal is to select a subset of objects to carry in the knapsack in such a way that their total value is maximum, subject to the constraint that their total weight does not exceed the knapsack's capacity.

If we look for an efficient algorithm for PARTITIONING we soon realize that a brute force approach is out of the question. Here too the space of all possible partitionings, given an input sequence of N values, has 2^N candidates (because every element has two choices, whether to stay in the first or in the second group) and therefore the trivial enumerative procedure is useless. The verification procedure, however, once again is rather straightforward and computationally inexpensive: given a partitioning of the input sequence, just compute two sums and check if they are the same.

The class NP contains all computational problems that are verifiable efficiently, like SAT and PARTITIONING. This is a very important class because it contains a myriad of computational problems of great practical interest. Indeed, NP problems are encountered in all fields where computers are used: biology, economics, all areas of engineering, statistics, chemistry, physics, mathematics, the social sciences, medicine and so on and so forth. One of the reasons why NP is so rich with problems of practical importance is due to the fact that it captures a huge variety of optimization problems. We already saw an example of an optimization problem, TSP: given the input (a map with N cities) we are seeking the optimal solution (the shortest tour) among a myriad of candidates implicitly defined by the input (all possible tours). Thus, in optimization we are seeking an optimal value inside a typically huge space of candidates. In contrast, NP is a class of decision problems, computational problems whose answer is either yes or no. There is however a standard way to transform an optimization problem into a decision problem. As an example, consider the following budgeted variant of TSP:

> **Budgeted TSP**
> **Input**: a map with N cities and a budget B (expressed, say, in kilometers)
> **Question**: Is there a tour whose length is no greater than B?

The problem asks if there is a way of visiting all the cities within the budget available to us, while in the original TSP we were looking for the tour of minimum length. Notice that now the answer is either yes or no. This budget trick is rather standard. With it we can turn any optimization problem into a decision problem. From a computational point of view, an optimization problem and its decision version are equivalent: if there exists an efficient algorithm to solve the optimization problem then there exists an efficient algorithm to solve the corresponding decision problem, and vice versa. In the case of TSP, for instance, if we solve the optimization problem by computing the shortest tour, then we can immediately answer the question *"Is there is a tour of length not exceeding the budget?"* thereby solving the decision problem. The converse is also true: If we can answer the question *"Is there is a tour of length not exceeding the budget?"* then we can efficiently determine the shortest tour (the slightly involved proof is beyond the scope of this book).

In order to appreciate the huge variety of contexts in which optimization problems arise it can be instructive to see one of the many examples taken from biology. DNA is a sequence of genes each of which is a set of instructions, a recipe, with which the cell can synthesize a specific protein. It is known that many species share the same gene pool, but the genes are positioned differently along the genome, i.e., genomes are anagrams of each other. For example a bonobo may have the following genome (letters represent genes):

PRIMATE.

While chimpanzees and humans might otherwise have, respectively, the genomes RPTIMEA and MIRPATE. Biologists postulate different mutation mechanisms that are responsible for the shuffling of genes during the course of evolution. One of these is the so-called "reversal". A reversal is due to a transcription error that literally inverts a segment of the genome, as in this example (the reversed part appears in bold):

PRIMATE → PRTAMIE

Biologists believe that reversals are rare events, taking place at intervals of millions of years. In our example the distance between bonobos and humans is only one reversal, **PRIM**ATE → **MIRP**ATE, while both species turn out to be farther away from chimpanzees (the reader can try to determine the minimum number of reversals between the three different species). Biologists assume that

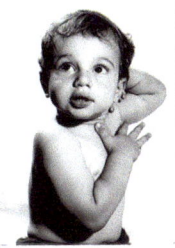

Fig. 3.2 From *left* to *right*: a young chimpanzee who already shows signs of aggressiveness, two bonobos in a normal loving attitude, and a lovely child

the minimum number of reversals to transform a genome into another is a good estimate of the evolutionary distance between the two species. Therefore, at least in this example, the lustful bonobo would be our closer relative than the irascible chimpanzee (Fig. 3.2). Thus, given two genomes, it is of interest to determine the minimum number of reversals between them. The corresponding decision version of this optimization problem is the following[9]:

Sorting by Reversals
Input: Two genomes G_1 and G_2 with the same set of genes, and a budget T.
Question: Is it possible to transform G_1 into G_2 with no more than T reversals?

The reader should try to convince him-/herself that this computational problem is in NP and that it is computationally equivalent to its optimization version (i.e., given two genomes find the minimum number of reversals that transforms one into the other).

In summary, the class NP captures a very large class of optimization problems of great practical importance. NP also contains other non-optimization problems that are of enormous industrial interest, such as the automatic verification of software and of robotic systems. But there is more. NP is also interesting from the point of view of the foundations of mathematics, a topic to which we now turn.

---

[9]The algorithmic structure of this important combinatorial problem was investigated by my late friend Alberto Caprara, who died tragically in a mountain accident. Alberto was able to establish that the problem is NP-complete (see next paragraph), thus settling an important open problem in the field of computational biology. The proof was quite a tour de force, and conveyed a sense of serene strength, quite like Alberto. This chapter is dedicated to him.

3.5 Universality

A beautiful and fundamental theorem concerning the class NP was established simultaneously and independently by Leonid Levin and Stephen A. Cook in the early 1970s. The theorem states that the class NP contains universal problems. A computational problem X in NP is universal (but the correct terminology is NP-complete) if it has the following property: if there is an efficient algorithm A for X, then, taken any other problem Y in NP (by means of a general procedure) we can transform algorithm A into an efficient algorithm B for Y. An immediate consequence of universality is: if we find an efficient algorithm for a universal problem then we can also find an efficient algorithm for every other problem in NP.

For example, satisfiability is universal: if one day you could find an efficient algorithm to decide whether a given Boolean formula is satisfiable, then it would be possible to efficiently solve any problem in NP. This result is known as Cook's Theorem (Levin had demonstrated the universality of a different problem, known as tessellation). Despite being a very strong property, universality appears to be the norm rather than the exception. Universal problems are plenty and we have encountered some of them already: satisfiability, partitioning, sorting by reversals and the budgeted versions of TSP. Other examples of universal problems are the decision versions of natural optimization problems like finding the longest path in a network, determining the smallest set of lecture halls to host university classes, packing a set of items with the smallest number of bins, and thousands more. For none of them it is known whether an efficient algorithm exists.

The concept of universality is very useful because it allows us to greatly simplify a complex picture without loss of generality. The class NP contains a bewildering variety of computational problems, how can we tell if there are efficient algorithms for all of them? Universality allows us to focus on just one problem, provided that it is universal. If we could find an efficient algorithm for a universal problem like TSP or satisfiability then we could efficiently solve any problem in NP. Conversely, if we could show that an efficient algorithm for TSP or satisfiability does not exist, then no universal problem may admit an efficient algorithm. In this way, the study of the "computational complexity", as it is called, of a myriad of problems can be reduced to the study of a single mathematical object.

3.6 The Class P

So far we have freely used the term "efficient algorithm", but what exactly do we mean by that?[10] The standard definition given in textbooks is the following: an algorithm is efficient if its computation time grows polynomially with the size of

---

[10] We revisit here concepts discussed in Chap. 2.

3 The One Million Dollars Problem

the input. It is intuitive that the larger the input the longer the calculation takes. For example, a multiplication algorithm will need more and more time to perform the product as the numbers to multiply become larger and larger. One of the missions of the theory of algorithms is to determine as precisely as possible the running time of algorithms, in a technology-independent manner (something that can be achieved thanks to appropriate mathematical models such as the Turing machine). If N is the number of bits needed to represent the input, the running time of the algorithm is "polynomial" if its computation time grows at most as a polynomial. For example, N, N^2, $N \log N$, $N^{3/2}$ are polynomial computation times ($\log N$ too is fine since it grows much slower than N). In contrast, we saw that the enumerative algorithms for SAT and TSP require exponential time, at least 2^N and $(N-1)!$ steps respectively, and we know that this kind of "complexity" is prohibitive, regardless of the technology.

NP is the class of problems whose solutions can be verified to be correct in polynomial time, while P is the class of problems for which the solution can be not only verified, but found in polynomial time. Various chapters of this book discuss important problems that are in P, such as the calculation of shortest paths in a network, the retrieval of information from the Web, and several fascinating examples taken from cryptology and computational biology. Other important computational tasks that are in P are the multiplication of matrices, solving systems of linear equations, determining whether a number is prime, the familiar "find" operation to look for the occurrence of a word inside a text, and the very general problem of linear programming (which asks us to optimize a linear function subject to linear constraints). The class P captures the set of all problems that are computationally tractable. We thus come to the most general formulation of the one million dollars problem:

$$P \stackrel{?}{=} NP$$

If $P = NP$ then every problem in NP can be solved efficiently, i.e., in polynomial time. In contrast, if $P \neq NP$, as it is widely conjectured, then no universal problem can be in P. (Why? The reader should try to answer.) If this is indeed the case we will have to accept our lot: we live in a world where computational intractability is the rule rather than the exception because, as we have discussed, NP contains a myriad of natural occurring computational problems that are universal (Fig. 3.3).

The theory of algorithms has thus developed two main approaches to ascertain the so-called "algorithmic complexity" of a problem. Given a computational problem, we can either exhibit a polynomial-time algorithm for it, or show that it is universal for NP (i.e., NP-complete). In the latter case we do not know for sure that the problem does not admit efficient algorithms, but we do know that finding one is a very difficult mathematical challenge, worth at least one million dollars! Let us stress once again that this theory is independent of technology: the watershed between exponential and polynomial complexity is such that no technological improvement will ever make exponential-time algorithms practical.

Fig. 3.3 The world as seen from NP. If P = NP then the three classes coincide, but if P ≠ NP then the rich class of Universal (NP-complete) problems is disjoint from P

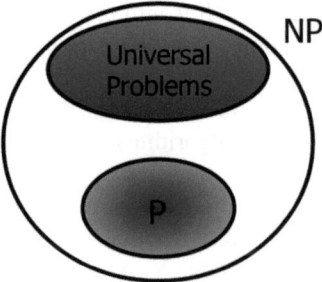

After having discussed the practical implications of the one million dollars problem, let us look at the other side of this precious coin, the link with the foundations of mathematics.

3.7 A Surprising Letter

As recalled, the theory of NP-completeness was developed in the early 1970s independently in the Soviet Union and in the West. An exceptional discovery made in the basement of the Institute for Advanced Study at Princeton in the late 1980s sheds new light on the history of the P vs. NP problem: it had been prefigured by no less than Kurt Gödel. In a letter that he wrote to John von Neumann, dated 20 March 1956, he states the P vs. NP problem quite ahead of its time. Although we do not know for sure, it is unlikely that von Neumann read the letter, or at least that he gave it much thought. At the time, unfortunately, he was lying in a hospital bed fighting the disease that would soon lead him to his death. In the letter, with a few quick, clean strokes Gödel goes back to Hilbert's Entscheidungsproblem arguing that, in a sense, his approach was simplistic. Suppose, Gödel says, that we find the proof of an interesting mathematical statement, but this proof is enormously long, say, it consists of $2^{1,000}$ symbols. This number, even when expressed in picoseconds, is well beyond the age of the universe. It is clear that such a proof would be of very little use: its length must be compatible with human life! The interesting question to ask therefore, is not so much whether a particular mathematical statement admits a proof but if a proof exists that can be grasped by mere mortals. To capture this more realistic situation in mathematical terms, consider the following decision problem, a sort of Entscheidungsproblem restricted to mathematical proofs that can be grasped by human beings[11]:

---

[11] We are sweeping a few technical issues under the carpet for the sake of simplicity. For a rigorous treatment see [13].

3 The One Million Dollars Problem

> **Existence of a Readable Proof (ERP)**
> **Input**: a mathematical statement E and a number N.
> **Question**: is there a proof of E of length at most N?

The question asked by Gödel in his letter is whether ERP admits an efficient algorithm: is it possible to find a readable proof, or conclude that none exists, in time polynomial in N, for example in time proportional to N or N^2? In this way Gödel identifies the true nature of the problem investigated by Hilbert and, embryonically if not naively, by Leibniz: can the creative process that leads to mathematical proofs be replaced by the work of a computer, i.e., a machine? To see why, let us compare the two approaches to this question, as proposed by Hilbert and by Gödel. Recall that Hilbert starts from the following computational problem (notice the difference with ERP):

> **Existence of a Proof (ED)**
> **Input**: a mathematical statement E.
> **Question**: is there a proof of E?

Let us now consider the following three questions, concerning these computational problems:

1. (Hilbert's Entscheidungsproblem) Is there an algorithm for ED?
2. Is there is an algorithm for ERP?
3. (Gödel's letter) Is there an efficient algorithm for ERP?

Gödel, perhaps because he was living at a time when electronic computers had become a reality, frames the problem correctly in a computational light. As he notes in his letter, an affirmative answer to the third question *"would have consequences of the utmost importance [as] it would imply that, in spite of the undecidability of the Entscheidungsproblem, the mental work of a mathematician with regard to yes/no questions could be replaced entirely by the work of a machine."*[12]

To clarify this point, suppose that there is an algorithm of complexity N^2 for the problem ERP and that we are interested in a particular statement E, e.g., the Riemann hypothesis. To find out if there is a proof of the Riemann hypothesis using ten million symbols, or about 350 pages, with current technology we would have to wait for a day or so. Possibly, something so deep and complex like the Riemann hypothesis might need more than 350 pages. No problem! By setting N equal to 100 million, corresponding to a proof of nearly 35,000 pages, you

---

[12] A translation of the letter can be found at http://blog.computationalcomplexity.org/2006/04/kurt-gdel-1906-1978.html

would have the answer within 4 years. It may seem a lot, but what about Fermat's Last Theorem, whose proof required nearly four centuries? To have the answer of what is considered the most important open problem in mathematics we might as well wait a few years! Moreover, technological improvements, providing ever more powerful hardware, would soon make manageable values of N over a billion, which is equivalent to considering arguments millions of pages long, a limit beyond which it certainly makes no sense to look for proofs.

We now note a trivial but surprising fact. The (negative) answer to question 1 required a flash of genius, in its most elegant form due to Turing. In contrast, the answer to question 2 is affirmative and rather trivial. Let us see why. Every mathematical theory can be described by means of standard mathematical formalism in which all the formulas, sentences, definitions, theorems and so on, can be written down by means of a finite set of symbols, the so-called alphabet (this should not be surprising, after all mathematics is written in books). Specifically, any mathematical theory uses the ten digits $0, 1, \ldots, 9$, uppercase and lowercase letters of the alphabet, sometimes Greek letters, special symbols like $\forall$, $\exists$, etc., parentheses (,), [,], and so on. Suppose, for the sake of simplicity but without affecting the generality of the argument, that our alphabet has 100 symbols. To determine whether a certain statement admits a proof with N symbols it is sufficient to generate all candidate proofs, i.e., all possible 100^N strings of symbols, and check them one by one. The checking part, to determine whether a given sequence of N symbols is a correct proof of a statement E, is doable by computer efficiently, in polynomial time.

The answer to question 2 therefore is affirmative: there is a simple, indeed trivial general procedure that allows a computer to tell, given any statement of mathematics whatsoever, whether there exists a proof understandable by human beings. For the proponents of human superiority over soulless machines this might sound like a death knell, but there is no need to worry. The by-now-familiar catch comes to our rescue: the algorithm that we outlined, having exponential complexity, is useless. To settle the question of human superiority once and for all we have to follow once again our lodestar – computational efficiency – and consider the third question, which asks whether machines can find understandable proofs efficiently. It should not come as a surprise by now to learn that ERP is universal (NP-complete). In other words, the problem posed by Gödel in his letter, question 3 above, which so effectively captures an old and formidable question – whether the creativity of a mathematician can be effectively replaced by a machine – is none other than the one million dollars problem:

$$P \stackrel{?}{=} NP$$

If $P = NP$, then, given any mathematical statement, a proof or a refutation can be found efficiently by an algorithm and, therefore, by a computer. It could be argued that in this case the discoverer of this exceptional algorithm could claim not one, but six million dollars, as it could answer all Millennium Problems remained unsolved until now![13] This seems very unlikely and is part of the circumstantial evidence that leads us to conjecture that the two classes must be different. This triumph of human creativity over the machines, however, would come at a heavy price, for we would have to resign ourselves to live in a world where, despite the impressive progress of technology, computational intractability remains the norm, even for trivial-looking tasks.

The Class P and Efficient Computations

A common, and quite sensible, objection that is raised when the class P is encountered for the first time is the following: if the running time of an algorithm is, say, proportional to $N^{1,000,000,000,000,000}$, we do have polynomial-time but this is as impractical as a running time like 2^N, and therefore P cannot be considered as a good characterization of the realm of feasible computations. To be sure, the intuitive concept of "feasible computation" is a very elusive one and P is only a raw characterization of it. It does capture however some of its fundamental aspects. A common occurrence is that, once a polynomial-time algorithm for a problem is found, however impractical, successive improvements bring about new algorithms of better and better performance. In practice, most polynomial-time algorithms have very low exponents. One could also argue that P embodies a conservative approach: if a problem does not have a polynomial-time algorithm, it must certainly be computationally unapproachable. The main point however is different and has to do with the combinatorial structure of algorithmic problems. As we have discussed, if we could afford exponential running times like 2^N, NP-complete problems like Satisfiability would admit trivial, brute-force algorithms: it would be sufficient to enumerate all possible candidate solutions (i.e., all possible truth assignments) and check them one by one, to see if any satisfies the formula. Suppose now that we could afford a running time of $N^{1,000,000,000,000,000}$. Even such a huge polynomial cannot match the exponential growth of 2^N and if we want an algorithm for SAT to cope with all possible inputs we cannot rely on a brute-force enumerative approach any more. In order to zero in on a satisfying assignment, or to conclude that none exists, a good deal of mathematical ingenuity would be needed to identify and exploit hidden organizational principles of the solution space (in the case of SAT, all possible truth assignments), provided that they exist. In fact, it is quite conceivable that no such regularities exist for NP-complete problems.

---

[13]Since the awards were established, the Poincaré Conjecture, one of the most coveted prey, has been answered affirmatively by Grigori Perelman. He refused, however, to collect both the prize and the Fields Medal!

3.8 The Driving Force of Scientific Discovery

"As long as a branch of science offers an abundance of problems, so long is it alive; a lack of problems foreshadows extinction" warned David Hilbert in his famous speech. Computer science certainly enjoys an extraordinary vitality in this respect. The pace of its fantastic technological achievements – a revolution that continues unabated in full swing – and the momentous social consequences that they engender, are perhaps the reason why the conceptual depth at the basis of this revolution is underappreciated. And yet, as we have seen, some of the scientific questions posed by computer science, of which P vs. NP is just one example, are of the greatest import, with ramifications deep and wide not only for mathematics but for many other disciplines. Nor could it be otherwise. Computer Science tries to understand the laws that govern two concepts that lie at the foundations of our understanding of the world: information and computation. These laws, which at present we are just beginning to glimpse, are as concrete as those of nature and, like those, define the boundaries of our universe and our ability to maneuver within it. The one million dollars problem is just the tip of a glittering iceberg.

3.9 Bibliographic Notes

The Millennium Problems and their relevance are discussed on the website of the Clay Mathematics Institute [79] (see Millennium Problems).

The theory of NP-completeness is now part of the standard university undergraduate curriculum in computer science around the world. Among the many excellent algorithms books now available I can certainly recommend "Algorithm Design" by Jon Kleinberg and Éva Tardos [67]. For a discussion from the viewpoint of computational complexity theory, a nice and agile book is "Computational Complexity" by Christos Papadimitriou [89]. The classic text, however, remains "Computers and Intractability: a Guide to the Theory of NP-Completeness" by Michael R. Garey and David S. Johnson [47]. This book played an important role in disseminating the theory, and it remains one of its best accounts.

An interesting article which describes the state of complexity theory in the Soviet Union is [106]. Kurt Gödel's letter to von Neumann is well analyzed in [13]. Both articles are available on the Web.

Turning to the popular science literature, a very nice account that explores the role of mathematical logic for the development of computers, from Leibniz to Turing, is "Engines of Logic: Mathematicians and the Origin of the Computer" by Martin Davis [24].

Biographies can offer interesting insights by placing ideas and personalities in their historical and cultural context, as well as reminding us that science, like all human activities, is done by people of flesh and bone, with their dreams, their pettiness and greatness, happiness and suffering. An excellent biography is "Alan

Turing: the Enigma" by Andrew Hodges [60], which tells the story of the tragic life of this great scientist, his role in decoding the secret code of the German U-boats during the Second World War, until the brutal persecution and breach of his fundamental human rights that he endured at the hands of the British authorities.

Finally, a wonderful booklet is "Gödel's Proof" by Nagel and Newman [83], a layman's masterly exposition of Gödel's magnificent Incompleteness Theorems.

Part II
The Difficult Simplicity of Daily Life

Chapter 4
The Quest for the Shortest Route

Camil Demetrescu and Giuseppe F. Italiano

Abstract Finding the shortest route is a ubiquitous problem in our daily life. Whenever we look for driving directions, surf the Web, send emails, or interact with our contacts in a social network, we are, perhaps unwittingly, exploiting the efficiency of an underlying shortest path algorithm. In this chapter, we review the basic ideas which are at the heart of shortest path algorithms and show how they seem to be related to some of the fundamental questions investigated by philosophers and scientists for centuries, in their effort to understand some of the deep mechanisms that rule the universe.

4.1 Introduction

While planning for a trip, whether it is a long-distance journey or just a short drive, we are often interested in taking the shortest possible route. This task can be automated with the aid of navigation systems, which deploy algorithms capable of finding the shortest, quickest or cheapest path among several possible alternatives. With today's impressive advances in technology, even small portable devices, such as smartphones and GPS navigation devices, are able to compute in a few seconds shortest routes on very large road networks covering many millions of roads and road junctions.

C. Demetrescu (✉)
Dipartimento di Ingegneria Informatica, Automatica e Gestionale, Sapienza Università di Roma, via Ariosto 25, 00185 Roma, Italy
e-mail: demetres@dis.uniroma1.it

G.F. Italiano
Dipartimento di Ingegneria Civile e Ingegneria Informatica, Università di Roma "Tor Vergata", via del Politecnico 1, 00133 Roma, Italy
e-mail: italiano@disp.uniroma2.it

Finding the shortest route appears as a subproblem in many other applications as well. Indeed, and perhaps surprisingly, most of our daily activities hinge on shortest paths, even when we are not traveling! As we are writing a piece of text or editing a file on a computer, the text editor is using a shortest path algorithm to format paragraphs, in order to balance the distribution of words within the same line or among different lines of the text. Whenever we search the Web, send email, chat, interact with our contacts in a social network or share files and other media (such as music, photos, and videos) over the Internet, we perform activities that are all based on shortest paths. In fact, when data is transferred on the Internet, deep down in the inner workings of computer networks each piece of data is chopped into small packets, which are then routed from their origin to their final destination by algorithms based on shortest paths. Similarly to road networks, in this case we can have several metrics for shortest paths, depending on whether the costs of the network links are given by their capacity, their congestion, or the physical distance between the corresponding endpoints.

Thus, the quest for the shortest route appears to be ubiquitous in our daily life. One might argue that this is due to the fact that, as human beings, we tend to transfer to computers our own problem-solving strategies in the form of sophisticated algorithms. On the other hand, one might wonder whether it is Nature itself which often tries to choose the shortest, simplest, or quickest way to achieve its goals. As a matter of fact, philosophers and scientists have been trying to address this question for centuries, in their effort to understand some of the deep mechanisms that rule the universe.

The anthropomorphic concept of a thrifty Nature, which operates by always choosing the most economical alternative, has indeed inspired many deep intuitions throughout the history of science. In the first century CE, Hero of Alexandria demonstrated that the assumption that light rays always travel between two points on the path of shortest length makes it possible to derive the law of reflection by using simple geometry. Sixteen centuries later, the French mathematician Pierre de Fermat (1601–1665) was able to push this idea even further, by postulating that "light travels through the path in which it can reach the destination in least time". This is Fermat's Principle, also known as The Principle of Least Time, and it is a fundamental law of optics from which the other laws of geometrical optics can be derived. After a few decades, Pierre-Louis de Maupertuis (1698–1759) extended the idea of a thrifty Nature from optics to mechanics, by introducing the Principle of Least Action, which basically states that Nature always finds the most efficient course from one point to another according to paths of least action, i.e., paths in which the total energy needed to get from one point to another is minimized. Since then, this paradigm has inspired several other important scientific developments, including the method of maxima and minima due to Leonhard Euler (1707–1783), and analytical mechanics due to Joseph-Louis Lagrange (1736–1813) and William R. Hamilton (1805–1865). Amazingly enough, the Principle of Least Action remains still central in modern physics and mathematics, and it has been applied in the theory of relativity, quantum mechanics, and quantum field theory.

4 The Quest for the Shortest Route

There are scents of shortest paths even among the oldest civilizations. According to Egyptian mythology, each day the sun-god Ra was born and began his journey across the sky. At sunset, the sky goddess Nut swallowed Ra, who passed through her body during the night, until he was born anew from her pelvis at the dawn of the new day. This beautiful image, which associates night with death and day with life or rebirth, reflects the typical Egyptian idea of immortality, and it can be taken as a symbol for the eternal cycle of life, death, and rebirth. In ancient Egypt, the myth of the solar child Ra and his mother Nut seemed to have close ties to actual astronomical observations. Indeed, the goddess Nut was frequently depicted as a personification of the Milky Way, a young goddess arching her body over the Earth, with her birth canal corresponding to Deneb, one of the stars of Cygnus. A line drawn from the North Celestial Pole through the star Deneb intercepts the horizon exactly at the point where the sun rises at Cairo every winter solstice: this line forms a great circle and represents the shortest path that the solar infant Ra would follow after exiting the birth canal to the point of appearance on the horizon at sunrise [109].

For ages, the shortest route appeared to be the main road for mortals and immortals alike. However, the shortest route may often contain hidden difficulties, and its quest can cost dearly people who ignore those hardships. The Carthaginian general Hannibal, for instance, on his march in the Roman territory in 217 BCE, planned to catch the Roman consul Gaius Flaminius off guard. To do this, he led his army to the mouth of the Sarnus river, a huge marsh which happened to be overflowing more than usual during that particular season. Hannibal knew that this way was full of difficulties. He could have taken a more comfortable path to central Italy, but decided to choose the shortest route instead. Paulus Orosius, a Christian historian from the fifth century CE, reports in his history the hardships incurred by Hannibal through the marshy lowlands of the Sarnus river [87]:

> So Hannibal, knowing that Flaminius, the consul, was alone in the camp, that he might more quickly crush him when unprepared, advancing in the early spring took the shorter but marshy road, and when the Sarnus happened to have overflowed its banks far and wide and had left the fields about it uncertain and loose, about which it has been said: "And the plains which Sarnus floods". When Hannibal proceeded into these fields with his army, with the mist especially as it rose from the marsh cutting off his view, he lost a large part of his allies and beasts of burden. He himself, moreover, seated upon an elephant which alone had survived, barely escaped the hardship of the journey; but he lost one eye, with which he had long been afflicted, because of the cold, lack of sleep, and hardships.

In 1474 Fernão Martins, canon at Lisbon Cathedral, was commissioned by King Alfonso V of Portugal to obtain some information from his friend Paolo dal Pozzo Toscanelli (1397–1482) as to the shortest route for reaching the East Indies by sailing to the west. This new curiosity was sparked by the fall of Constantinople to the Ottoman Turks in 1453, which make the land route to Asia more difficult, and forced western Europe to look for new trade routes between Europe and Asia. Toscanelli was considered to be one of the best mathematicians of his time, and he was also an astronomer and a cosmographer. His reply to Martins detailed a scheme for sailing westwards to reach the Spice Islands and Asia, and argued that the distance from Lisbon to Cathay (China) was only 6,500 miles, which were supposed to be one third of the earth's circumference:

> Paul, the physicist, to Fernando Martinez, canon, at Lisbon, greeting. [...] I have formerly spoken with you about a shorter route to the places of Spices by ocean navigation than that which you are pursuing by Guinea. [...] in order to make the point clearer and to facilitate the enterprise, I have decided to exhibit that route by means of a sailing chart. I therefore send to His Majesty a chart made by my own hands, upon which are laid down your coasts, and the islands from which you must begin to shape your course steadily westward, [...] From the city of Lisbon due west there are 26 spaces marked on the map, each of which contains 250 miles, as far as the very great and splendid city of Quinsay. [...]

The King of Portugal did not follow Toscanelli's suggestions and decided to pursue the alternative route through Southern Africa. In 1474, however, Columbus requested the same information from Toscanelli, and received back a copy of the letter he had sent to Martins, which Columbus transcribed in his copy of Piccolomini's Historia, accompanied again by a chart. Toscanelli's claims about the distance from Lisbon westwards to China were not correct, and Columbus himself added more errors by considering a worse estimate of the earth's circumference. As a result, he believed that the distance between Spain and Japan was only about 3,000 nautical miles, rather than the actual 10,600 nautical miles. Despite those errors, Columbus embarked on what he believed to be the shortest sea route to the Indies, and he ended up discovering America by mistake.

4.2 The Mathematisch Centrum

The idea of promoting the application of mathematics in the Netherlands was conceived during World War II. The Mathematisch Centrum (MC), now Centrum voor Wiskunde en Informatica (CWI), opened in Amsterdam in 1946, and many institutions, including the Philips Gloeilampen Fabrieken (now Philips Electronics), contributed to its foundation. One of the first decisions taken was to build a computer, named ARRA, which was completed after 6 years in 1952. Unfortunately, the ARRA never worked satisfactorily and it was broken up shortly after its official inauguration. In the same year, 22-year-old Edsger W. Dijkstra (1930–2002) joined the Mathematisch Centrum. Dijkstra was a brilliant student in the University of Leiden, who was interested in theoretical physics. In 1951 his father saw an advertisement for a 3-week summer course in computer programming at the University of Cambridge, and he advised his son to attend the course. Feeling that programming was a good skill to have for a theoretical physicist, the young Edsger took the course, a decision that would eventually change his life. Aad van Wijngaarden, Director of the Computation Department of MC, had taken the same course in the previous year; when he learned that Dijkstra had completed the summer course in Cambridge he decided to hire him as a programmer to write the basic software for ARRA II, the next MC computer. Dijkstra accepted the position but only as a part-time job, since he wanted to complete his studies in theoretical physics at the University of Leiden. The ARRA II was completed in 1955 and worked successfully, thanks also to the influence of Gerrit A. Blaauw (1924–), who had gained some experience in the field by working with Howard Aiken (1900–1973) at IBM.

Dijkstra graduated in theoretical physics in 1956 and moved to Amsterdam to work full-time at MC. In the same year, the third MC computer, named ARMAC, was ready for use, with much of its basic software being programmed by Dijkstra. For the celebration of the ARMAC inauguration, Dijkstra was asked to prepare a demonstration to show the power of the new computer. It was crucial to find a fundamental problem whose importance could be easily understood even by non-specialists. Could anything have been better than the ubiquitous quest for the shortest route? As Dijkstra said in a later interview:

> [...] for a demonstration for non-computing people you have to have a problem statement that non-mathematicians can understand, even they have to understand the answer. So I designed a program that would find the shortest route between two cities in the Netherlands, using a somewhat reduced roadmap of the Netherlands, on which I had selected 64 cities (so that in the coding, 6 bits would suffice to identify a city).

This is how Dijkstra described that important day in one of his papers [32]:

> For the purpose of the demonstration, I drew a slightly simplified map of the Dutch railroad system, someone in the audience could ask for the shortest connection between, say, Harlingen and Maastricht, and the ARMAC would print out the shortest route town by town. The demonstration was a great success; I remember that I could show that the inversion of source and destination could influence the computation time required. The speed of the ARMAC and the size of the map were such that one-minute computations always sufficed.

To solve this problem, Dijkstra invented a new algorithm for computing efficiently the shortest paths. First of all, it was necessary to formalize the problem in a proper mathematical setting, and for this graph theory came in handy.

4.3 Shortest Paths in Graphs

We can think of a railway system in abstract terms, as was done by Dijkstra, by using a graph-theoretical approach. Graphs are fundamental mathematical structures that are used to model relations between objects from a certain collection. In particular, a graph consists of a set of *vertices* and a set of *edges* that connect pairs of vertices. In this context, each city in a railway network can be represented by a vertex, and there is an edge between two vertices if there is a direct rail line between the two corresponding cities. Each train route will therefore be associated to a suitable sequence of graph vertices which represent the cities crossed by that route, and a shortest path between two cities in the railway network will correspond to a shortest path in the graph. This correspondence allows us to abstract the essence of the problem mathematically, so that it can be possible to solve it with a computer program.

What do we mean exactly by *shortest* path? It clearly depends on what we are interested in. For instance, we might be interested in computing the train route with the smallest number of city stops. In this case, we have to find the path with the smallest number of edges in the corresponding graph. On the other hand, we might want to minimize the geometric distance traveled, the travel time, or the total rail fare for our journey. Of course, different criteria will yield pretty much different results. This must have been clear to Hannibal, since taking the shorter but marshy road close to the Sarnus river resulted in a much higher cost overall!

How can we add information about geometric distances, travel times or rail fares to the abstract representation of a graph? This can be done by associating with each edge in the graph a real-valued cost, which represents the length, the travel time, or the fare of the corresponding rail line. In this framework, the cost of a path will be given by the sum of the costs of its edges, and the shortest path between two vertices in the graph will give respectively the shortest, quickest or cheapest train route between the two corresponding cities. We remark that the edge costs play a crucial role in the correct formulation of the problem. This could be witnessed by Columbus: after assigning the wrong cost (3,000 rather than 10,600 nautical miles) to an edge, he sailed towards the Indies on a sea route quite different from the shortest path!

4.4 Nature and Its Algorithms

We now go back to our original question: is Nature able to find the shortest route? To answer that question, we look at a graph from a geometric viewpoint. Towards this aim, we realize a "physical" graph with balls and strings, as illustrated in Fig. 4.1. We put on a table a ball for each vertex u in the graph. For each edge (u, v) we cut a piece of string, with length proportional to the cost of edge (u, v), and tie the string's endpoints to the two balls representing vertices u and v, respectively. The result is depicted in Fig. 4.1b: the mathematical structure of the graph is simply represented by a set of balls and strings!

Now take two specific balls: if you pull them gently far apart, some strings will be in tension until you reach a point where pulling the two balls any further will break one of the strings (see Fig. 4.1c). It is not difficult to see that the set of strings in tension between the two chosen balls represents a shortest path between the corresponding vertices. In this case, Nature is able to find the shortest route!

We now perform another experiment, but this time we use flammable balls and replace each string with a line of gunpowder. We assume that flames travel at the same speed down the line: i.e., if we take two lines of gunpowder having the same length and light them up at the very same time, they will burn completely in the same amount of time. If we now light up a given ball s, then the fire front will get simultaneously to all lines of gunpowder connected to s.

What is the second ball that will burn after s? Obviously, the ball which has the shortest line of gunpowder to s. The fire front will attack balls and lines of

4 The Quest for the Shortest Route

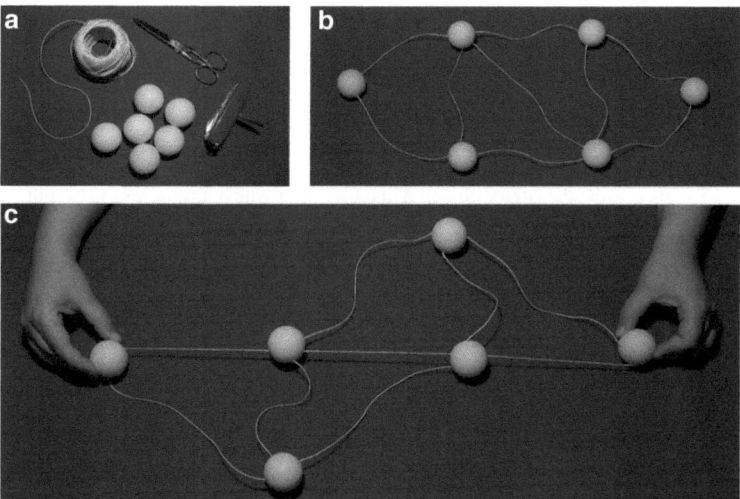

Fig. 4.1 (a) Ingredients and tools used for realizing a graph: balls, strings, scissors and a Swiss army knife. (b) A graph with six vertices and nine edges built with balls and strings. (c) A shortest path is given by a sequence of strings in tension

gunpowder until all of the graph is completely burnt out. The time at which a particular ball v burns will be proportional to the distance from s to v, i.e., to the length of the shortest path from s to v. Furthermore, the predecessor of v in this shortest path is given by the line of gunpowder (u, v) which caused v to light up. Once again, Nature computes shortest routes!

4.5 A Simple Idea

Dijkstra thought about his algorithm on a sunny morning in 1956, while drinking coffee with his wife in a cafe terrace in Amsterdam. To find the shortest path between two vertices, he considered the more general problem of finding the shortest path from a given source vertex to all other vertices. As we saw in the previous chapters, an algorithm is defined by a finite sequence of basic operations, which a particular executor (such as the CPU of a computer) is able to perform in order to produce a final result starting from some input data. In our setting, the input is given by a weighted graph and a starting vertex s. The final result that we would like to obtain as output is, for any vertex v, its *distance* from vertex s (denoted by $d(v)$) and its *predecessor* in a shortest path from s to v (denoted by $p(v)$). As we will see next, once all distances $d(v)$ are available, then also the predecessors $p(v)$ can be easily computed. In the following, we will thus restrict ourselves only to the computation of the distances $d(v)$.

> **Input:** A weighted graph with vertex set V and a source vertex s
> **Output:** For each vertex v, its distance $d(v)$ from s
>
> **Algorithm Basic Dijkstra**
>   **Step 1:** Let A be the set consisting of vertex s and let $d(s) = 0$
>   **Step 2:** Repeat the following steps until $A = V$:
>   **2.1** Let (u,v) be the edge with u in A and v not in A which minimizes the quantity
> $d(u) +$ cost of (u, v)
>   **2.2** Set $d(v) = d(u) +$ cost of (u,v)
>   **2.3** Add v to set A

Fig. 4.2 Dijkstra's algorithm

Dijkstra's algorithm is a classic example of how a mathematical property can yield a simple and elegant procedure. Assume that all the distances from source s to a set A of vertices have been already computed, and let (u, v) be the edge, having only one end in A, that minimizes the quantity $d(u) +$ cost of (u, v). Dijkstra observed that this edge (u, v) is particularly important, as it allows one to compute the distance from s to v:

$$d(v) = d(u) + \text{cost of } (u, v)$$

As a consequence of this property, if we find the proper edge (u, v), we can find one more vertex v for which we know the exact distance from s! This step can be repeated, until we find the distances from vertex s to all other vertices. Initially, we can start from $A = \{s\}$, since we only know the exact distance from s to itself ($d(s) = 0$). Figure 4.2 summarizes the steps of this simple algorithm.

There is a striking analogy between the algorithm given in Fig. 4.2 and the process of fire propagation illustrated in Sect. 4.4. In this analogy, the vertex set A corresponds to the balls that already caught fire, and the edges with exactly one endpoint in A correspond to the lines of gunpowder that are currently burning. The edge (u, v) chosen in Step 2.1 of Fig. 4.2 is the next line of gunpowder that will burn completely, and the vertex v will correspond to the ball that will catch fire next. Dijkstra's algorithm can thus be seen as a discrete simulation of the continuous combustion process that burns balls and lines of gunpowder!

The algorithm of Fig. 4.2 computes only distances. However, as was mentioned before, it can be easily modified to keep track of all predecessors $p(v)$. Indeed, if we set $d(v) = d(u) +$ cost of (u, v) on Step 2.2 of Fig. 4.2, then the shortest path from s to v is given by the shortest path from s to u followed by edge (u, v), and thus vertex u is the predecessor of v along the shortest path from s. Consequently, to compute all predecessors it is enough to set $p(v) = u$ on Step 2.2. Figure 4.3 shows an example of how Dijkstra's algorithm computes all distances from a given source vertex.

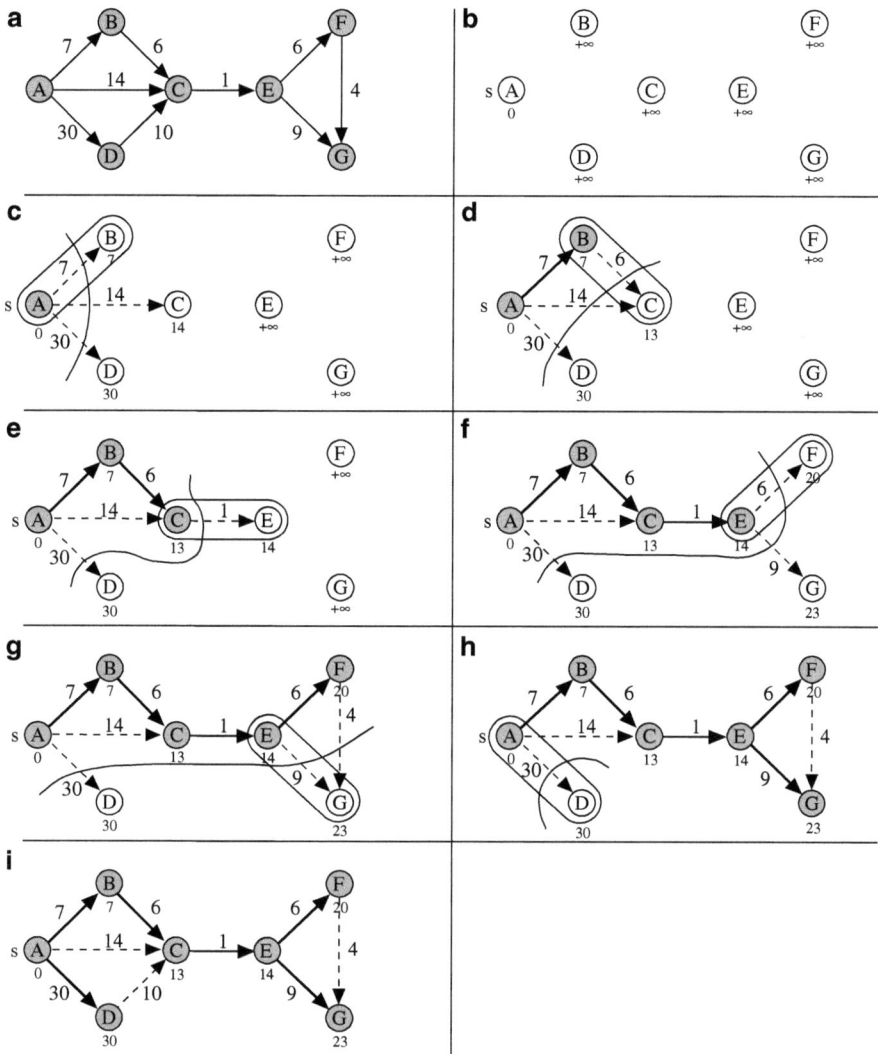

Fig. 4.3 (a) A weighted graph; (b)–(i) the steps of Dijkstra's algorithm on the graph in (a) with source vertex A. At each step, the tentative distance of each vertex from v is shown below the vertex itself. Initially (b), the source vertex has distance 0 from itself, and the remaining vertices have tentative distance equal to $+\infty$. In the following steps (c)–(i), actual distances are computed with the formula $d(v) = d(u) + \text{cost of } (u, v)$. At each step, all the vertices in A are shown *gray*, and the chosen edge is highlighted with a *box*

4.6 Time Is a Tyrant

How fast is Dijkstra's algorithm? Of course, it depends on the size of the input graph. Let m and n denote, respectively, the number of edges and vertices in the graph. The bottleneck of the algorithm seems to be the search for the edge (u, v) in Step 2.1. This step is repeated at most n times, since each time we execute Step 2.1, a new vertex gets added to the set A in Step 2.3. A naive implementation Step 2.1 requires us to scan all edges in the graph, and thus it can take as many as m basic operations (such as sums and numeric comparisons). This implies that Dijkstra's algorithm can be implemented with a total of $m \times n$ basic operations in the worst case.

Such an algorithm would be rather unpractical on large-scale graphs. Assume that we wish to plan a car trip from Rome to Amsterdam, and are interested in finding the shortest route using the trivial implementation of Dijkstra's algorithm described before. The road network of Western Europe is a graph with roughly 18 million vertices and 42 million edges, where vertices are intersections and edges are road segments between intersections, with their cost being the average travel times [93]. On a graph with $n = 1.8 \times 10^7$ vertices and $m = 4.2 \times 10^7$ edges, finding the shortest path requires $m \times n = 7.5 \times 10^{14}$ basic operations, i.e., 750 trillion basic operations! Even on today's computing platforms, with CPUs clocked at about 3 GHz, this would imply several months of computing time! If this computation had started in 1956 on the ARMAC computer, it would still be going on today!

We now show how this algorithm can be sped up. As we saw before, the bottleneck of the algorithm is finding the next vertex v to be added to the set A. This is trivially accomplished in Step 2.1 of Fig. 4.2 by looking for the edge (u, v) that has exactly one endpoint in A and that minimizes the quantity $d(u) +$ cost of (u, v). Can we check this faster than Step 2.1? Going back to our fire analogy, if at any time we take any snapshot of our burning graph, the set A corresponds to balls that already burned. We can partition the remaining balls into two disjoint sets, denoted by B and C, as follows: B contains all unburned balls that are connected to at least one burning line of gunpowder, while C contains all the other unburned balls (i.e., unburned balls which are connected to no burning line). Note that the set B includes the next ball v that will catch fire, which is exactly what we would like to compute more efficiently. In graph-theoretical terms, B contains the vertices that are not in A but are connected directly through an edge to vertices in A, and C contains all the remaining vertices (which are neither in A nor in B).

For each vertex y in B, consider the edge (x, y), with x in A, which minimizes the quantity $d(x) +$ cost of (x, y), and denote this quantity by $D(y)$. We can think of $D(y)$ as the earliest time at which the fire coming from a line of gunpowder with one end in A will reach ball y. With this approach, the next vertex v to be added to set A is the one with the smallest possible value of $D(v)$. Put in other words, if $D(y)$ is available for all vertices y in B, to find the next vertex v to be added to A we do not need to scan all the edges with exactly one endpoint in A, which can be as many as m, but only the vertices in B, which are at most n. Since m can be as

> **Input:** A weighted graph with vertex set V and a source vertex s
> **Output:** For each vertex v, its distance $d(v)$ from s
>
> **Algorithm Faster Dijkstra**
>   **Step 1:** Let B be the set consisting of vertex s,
>   let $D(s) = 0$ and let $D(x) = +\infty$ for each vertex $x \neq s$
>   **Step 2:** Repeat the following steps until $B = \emptyset$:
>   **2.1** Let v be the vertex in B with smallest $D(v)$
>   **2.2** Remove vertex v from B
>   **2.3** Set $d(v)$ to $D(v)$
>   **2.4** For each edge (v,z) incident to vertex v do the following:
>   **2.4.1** If $d(v)$ + cost of (v,z) is smaller then $D(z)$ then
>   **2.4.1.1** Add vertex z to B (unless z is not already in B)
>   **2.4.1.2** Set $D(z) = d(v)$ + cost of (v, z)

Fig. 4.4 A faster version of Dijkstra's algorithm

large as about n^2 (there can be one edge between any vertex pair), we expect this observation to produce a faster implementation of Dijkstra's algorithm.

The resulting algorithm, which is closer to Dijkstra's original formulation, is illustrated in Fig. 4.4. The basic idea behind this algorithm is to maintain for each vertex x the value $D(x)$ defined previously. More precisely, $D(x)$ is an upper bound on the distance $d(x)$ from the source vertex s, i.e., $D(x) \geq d(x)$, and we refer to $D(x)$ as the *tentative distance* of x from s. Note that a trivial upper bound of the distance $d(x)$ is always $D(x) = +\infty$, and the goal of the algorithm is to compute the best possible estimate $D(x) = d(x)$ for each vertex x. Throughout its execution, the values $D()$ are used by the algorithm to identify the three disjoint vertex sets A, B and C, as follows.

1. For each vertex x in A, $d(x)$ has been already computed and $D(x) = d(x)$: the corresponding ball x burned already, and its tentative distance $D(x)$ equals its actual distance $d(x)$.
2. For each vertex y in B, $d(y) \leq D(y) < +\infty$: vertex y has a nontrivial (i.e., finite) tentative distance $D(y)$, and the corresponding ball y is connected directly to the fire front; $D(y)$ gives the earliest time at which the fire will arrive in y directly through a currently burning line.
3. For each vertex z in C, $D(z) = +\infty$: the corresponding ball z did not catch fire yet, and it is not connected to a burning line, so we can only have a trivial estimate (i.e., $D(z) = +\infty$) of when z will catch fire.

We remark that the algorithm in Fig. 4.4 maintains explicitly only the set B, as the two other sets A and C can be identified implicitly from the values of the tentative distances $D()$.

Initially (i.e., before starting the fire from ball s), we only know that $D(s) = 0$ (i.e., the source s will catch fire at time $t = 0$), and that $D(x) = +\infty$ for $x \neq s$. Thus, $B = \{s\}$. There is no fire front yet, and thus no other ball is connected to the fire front with a burning line. At the generic iteration, the algorithm selects the vertex v in B with smallest tentative distance (Step 2.1), i.e., it identifies the unburned ball v that will catch fire next. When the ball v will go off, the lines of gunpowder (v, z) (not yet burned) connected to v will catch fire, as considered in Step 2.4. Each line (v, z) will contribute to the computation of the possibly new minimum value for $d(v)$ + cost of (v, z) that defines $D(z)$ (Steps 2.4.1 and 2.4.1.2) and to the update of the set B (Step 2.4.1.1).

What is the running time of this algorithm? Since each vertex can be inserted and removed from the set B only once, Steps 2.1–2.3 and Step 2.4.1.1 are executed at most n times each. Note that Steps 2.2–2.3 and Step 2.4.1.1 requires a constant number of basic operations, while Step 2.1 requires us to find the minimum in B. The minimum in B can be computed by simply scanning all the elements of B, and thus in at most n basic operations. On the other hand, the remaining steps (i.e., Steps 2.4, 2.4.1 and 2.4.1.2) are executed at most once for each edge, and thus at most m times each. As a result, the algorithm in Fig. 4.4 can be implemented in approximately $n^2 + m$ basic operations, if n and m are, respectively, the number of vertices and edges in the input graph.

Is this a significant improvement? We observe that $n^2 + m$ is better than $m \times n$ whenever m is much larger than n. When m is close to n, however, $n^2 + m$ is not much better than $m \times n$. On the road network of Western Europe, with approximately 18 million vertices and 42 million edges, this improved version of Dijkstra's algorithm would still require about $n^2 + m = 3.2 \times 10^{14}$, i.e., roughly 320 trillion basic operations, which is only slightly better than the $m \times n = 7.5 \times 10^{14} = 750$ trillion basic operations required by the first version of the algorithm illustrated in Fig. 4.2. Thus, this improved version of Dijkstra's algorithm would still require months of computations on today's computing platforms. In 1956, Dijkstra's demonstration took only few minutes on the ARMAC, but on a much smaller graph with only tens of vertices and edges. Nevertheless, today we are able to compute in few seconds shortest paths on graphs with million vertices and edges using Dijkstra's algorithm. How can that be possible?

4.7 How to Set Your Priorities

One of the crucial (and most expensive) tasks of Dijkstra's algorithm is to select the element of minimum value in the set B (Step 2.1 of Fig. 4.4). We call the value $D(v)$ of each element v its *priority*. Note that the set B is changing dynamically, as new elements may be added to B (Step 2.4.1.1 of Fig. 4.4), removed from B (Step 2.2 of Fig. 4.4) or have their priority decreased (Step 2.4.1.2 of Fig. 4.4). In other terms, the algorithm needs to maintain a data structure capable of performing the following operations:

- Add a new item x with priority $D(x)$ to the set B (operation *insert*);
- Decrease the priority of item x to a new value $D(x)$ (operation *decrease*);

4 The Quest for the Shortest Route

Table 4.1 Faster implementations of the heap data structure and their impact on the running times of Dijkstra's algorithm. For each heap operation, the corresponding column shows the number of basic operations required by the operation on average, up to multiplicative factors. The last column shows the total number of basic operations required by Dijkstra's algorithm with the different heap implementations considered, again up to multiplicative factors

Year	Insert	Delete	Find minimum	Decrease	Running time of Dijkstra's algorithm
1956	1	1	n	1	$n^2 + m$
1964	$\log n$	$\log n$	1	$\log n$	$(n+m) \cdot \log n$
1987	1	$\log n$	1	1	$m + n \cdot \log n$

- Find the item of minimum priority in B (operation *find minimum*);
- Remove the item x from set B (operation *delete*).

This data structure is known as a *heap* (or alternatively *priority queue*). Heaps were not well known in the 1950s, when Dijkstra first designed his shortest path algorithm. However, today they are ubiquitous data structures in many computations. If we run Dijkstra's algorithm of Fig. 4.4 on a graph with m edges and n vertices, and maintain the set B as a heap, then it is possible to see that we will have to perform at most n insert, delete or find minimum operations. This is a consequence of the fact that, once removed from B, a vertex cannot be reinserted back, and thus each vertex can be inserted and removed only once from B. On the other side, if the priority of a vertex v is decreased then we are scanning an edge to v, and thus the total number of decrease operations will be proportional to m in the worst case. If we denote by $t_{\text{find minimum}}$, t_{insert}, t_{delete} and t_{decrease}, respectively, the running times of the find minimum, insert, delete and decrease operations, then the total running time of Dijkstra's algorithm will be proportional to

$$n \cdot (t_{\text{find minimum}} + t_{\text{insert}} + t_{\text{delete}}) + m \cdot t_{\text{decrease}}.$$

As a consequence, the performance of Dijkstra's algorithm depends on the particular heap deployed. Although the key algorithmic idea conceived in 1956 did not change throughout the years, the dramatic progress on the design of efficient heap data structures has produced faster solutions to the shortest path problem. Table 4.1 reports some of the milestones in the progress of efficient heap implementations, and their impact in Dijkstra's algorithm. The program written for the ARMAC in 1956 implemented the heap as a simple linked list: within this framework, inserting, removing and decreasing the priority of an item could be done in a constant number of steps, but finding the minimum required that we scan the entire list, and thus a number of basic operations proportional to n. This is illustrated in the first row of Table 4.1; we will next see more sophisticated implementations of heaps, capable of achieving the bounds contained in the last two rows of Table 4.1.

In 1964, John W. J. Williams [110] designed an ingenious method to implement heaps: the number of basic steps required to find the minimum would improve from n to constant, at the price of increasing from constant to $\log_2 n$ the basic steps

Fig. 4.5 A heap with six nodes

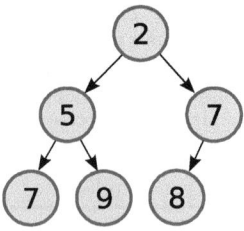

required by the other operations. The interested reader is referred to the next section for the details of the method, which accounts for the second row in Table 4.1.

4.7.1 The Heap Data Structure

The basic idea of Williams was to organize the data of a heap in a binary tree (see Chap. 2), with the following three additional properties:

Property 1: Each element in the heap corresponds to a binary tree node (in the following, we will use interchangeably the terms heap element and tree node).
Property 2 (heap property): Each non-root node is greater than or equal to its parent.
Property 3 (shape property): All levels of the tree, except possibly for the last one (deepest) are fully filled; if the last tree level is not complete, the nodes of that level are filled from left to right.

Figure 4.5 illustrates an example of a heap. Given n elements, Property 1 guarantees that the heap contains exactly n nodes. As a consequence of Property 2, all nodes are greater than or equal to the tree root, and thus operation *find minimum* can be implemented by simply returning the tree root. Finally, Property 3 can be used to prove that the height of a heap is proportional to $\log_2 n$. We now show how to implement efficiently the other operations *insert*, *delete* and *decrease*. In particular, we will see how the number of basic steps required to support each operation is proportional to the tree height, i.e., to $\log_2 n$.

Consider operation *decrease* first. After decreasing the priority of a node v, then Properties 1 and 3 will trivially hold. However, if the new priority of v is less than the priority of the parent of v, then Property 2 might be violated. In this case, we can restore Property 2 by repeatedly swapping v with its parent until they are in the correct order. This can be accomplished with the algorithm *up heap* illustrated in Fig. 4.6. Since the height of a heap with n nodes is proportional to $\log_2 n$, then *up heap* and consequently *decrease* can be implemented with $\log_2 n$ basic steps in the worst case.

To perform operation *insert*, we first add the new node to the bottom level of the heap, so that the shape property is maintained. Similarly to the previous case, Properties 1–3 hold except for the newly inserted node, which may violate

> **Input:** A binary tree for which Properties 1–3 hold, except for node v, which violates Property 2
> (with v being smaller than its parent)
> **Output:** The modified tree so that it is a heap
>
> **Algorithm up heap**
>   Repeat the following step
>   Swap v with its parent
>   Until either v is the tree root or v is greater than or equal to its parent

Fig. 4.6 Algorithm *up heap* restores Property 2 by repeatedly swapping a node with its parent until they are in the correct order

> **Input:** A binary tree for which Properties 1–3 hold, except for node v, which violates Property 2
> (with v being larger than its smaller child)
> **Output:** The modified tree so that it is a heap
>
> **Algorithm down heap**
>   Repeat the following step
>   Swap v with its smaller child
>   Until either v is a leaf or v is less than or equal to its children

Fig. 4.7 Algorithm *down heap* restores Property 2 by repeatedly swapping a node with its smaller child until they are in the correct order

Property 2. We can thus restore this property by applying again the algorithm *up heap* of Fig. 4.6 starting from the new node. Again, this will require $\log_2 n$ basic steps in the worst case.

It remains to show how to implement operation *delete*. Let u be the node to be deleted from the heap. Note that, unless u is the last node on the last level (i.e., the rightmost leaf in the tree), the deletion of u might cause a violation of Property 3 (shape property). To maintain this property, we replace the node u to be deleted with the rightmost leaf. At this point, we have restored Properties 1–3, except for the replaced node, which might now violate Property 2. If the replaced node is smaller than its parent, then it must move up the tree. Otherwise, if the replace node is larger than either of its children, it must move down the heap. The former case can be dealt with the algorithm *up heap* of Fig. 4.6. In the latter case we can repeatedly swap the node with the smaller child, until they are in the correct order. This can be accomplished with the algorithm *down heap* described in Fig. 4.7, which again requires time proportional to the height of the heap (i.e., $\log_2 n$ basic steps).

In summary, with this approach a *find minimum* requires constant time, while the remaining heap operations (*insert*, *delete* and *decrease*) can be implemented in approximately $\log_2 n$ basic steps each. Plugging this data structure into Dijkstra's algorithm yields a $(m+n) \log_2 n$ bound, as illustrated in the second row of Table 4.1. Two decades later, in the late 1980s, Michael L. Fredman and Robert E. Tarjan contributed another major step to the shortest path problem. Indeed in a famous article published in a scientific journal in 1987 [45], Fredman and Tarjan presented a new heap, called the *Fibonacci heap*. The name comes from Fibonacci numbers, which are used in the running time analysis of this data structure. The improvement of Fibonacci heaps over traditional heaps is that they are able to support *insert* and *decrease* operations faster in average constant time rather than logarithmic time. Dijkstra's algorithm implemented with Fibonacci heaps requires $m + n \log_2 n$ basic steps (improved from the $(m + n) \log_2 n$ bound obtained with Williams' classic heaps). On our road network of Western Europe, with approximately 18 million vertices and 42 million edges, these advances in data structure technology bring the running time for computing a shortest path with Dijkstra's algorithm from several months down to a few seconds!

4.8 The Humble Programmer

Let us go back to 1956. Although Dijkstra was fully aware of the importance of his discovery, he did not publish his result until 1959 [33]:

> At the time, algorithms were hardly considered a scientific topic. I wouldn't have known where to publish it.... The mathematical culture of the day was very much identified with the continuum and infinity. Could a finite discrete problem be of any interest? The number of paths from here to there on a finite graph is finite; each path is a finite length; you must search for the minimum of a finite set. Any finite set has a minimum – next problem, please. It was not considered mathematically respectable...

Dijkstra's algorithm was published only 3 years after its discovery, and it is one of the most cited and celebrated articles in computer science. Dijkstra himself recalled the birth of that three-page article, which was published in 1959 in the first issue of *Numerische Mathematik*, the first journal devoted to automatic computing [32]:

> In retrospect, it seems strange that I waited another two years before I submitted them for publication. The main reason was the absence of journals dedicated to automatic computing, something the foundation of *Numerische Mathematik* sought to remedy. I wrote and submitted my little article – my second one – trying to assist the fledgling. Furthermore, the publish-or-perish syndrome had not reached the Netherlands yet.

Dijkstra's algorithm was not known for several years after its publication. Indeed, the classic 1962 book by Lester Ford and Delbert Fulkerson [44] on network optimization cited many other inferior algorithms. This explains why Dijkstra's algorithm was later rediscovered by several other researchers, George Dantzig included. In his Turing Award Lecture [31], Dijkstra called himself a "humble programmer", and recalled the times where he was hired as a programmer at the Mathematisch Centrum, where he designed his famous shortest path algorithm, which was only intended for a demo.

> Another two years later, in 1957, I married and Dutch marriage rites require you to state your profession and I stated that I was a programmer. But the municipal authorities of the

town of Amsterdam did not accept it on the grounds that there was no such profession. And, believe it or not, but under the heading "profession" my marriage act shows the ridiculous entry "theoretical physicist"!

4.9 Still an Open Challenge

As was described earlier, Dijkstra's algorithm is able to compute all the shortest paths from a given vertex to all other vertices in the graph, a problem known in the scientific literature as the *single-source shortest path*. However, the demonstration prepared for the inauguration of the ARMAC computer was intended to solve an apparently simpler problem, namely the *single-pair shortest path*, which consists of finding a shortest path between two given vertices. In the last decade, many researchers investigated the simpler single-pair shortest path problem and tried to design algorithms that require time proportional to the size of the shortest path rather than to the size of the entire graph.

An efficient solution to the single-pair shortest path problem appears indeed of great practical importance. Consider, for instance, GPS navigation systems, which typically run on low-end architectures, including smart phones and small portable devices. The goal of a navigation system is to compute the shortest/fastest route between any two given points of a road network. A navigation system must be able to answer queries in reasonable times on road networks, which typically have millions of vertices and edges. Thus, it cannot afford to explore the entire road map, but rather it can only visit a smaller portion of the map. Back in 1956, Dijkstra used the following simple heuristic to speed up the computation of his program: as soon as the final destination was reached, the algorithm would be stopped. This might not be always sufficient, however, as the algorithm could still explore a large portion of the graph before reaching the final destination. Figure 4.8a shows the result of an experiment on a portion of the North American road network around Seattle, having about 1.6 million vertices and 3.8 million edges. The vertices explored by Dijkstra's algorithm before reaching the final destination are shown in dark gray, while the shortest path is shown in black. As can be clearly seen from the figure, the dark gray region covers almost the entire road network, while the shortest path contains only few hundred vertices.

A faster method could be to run a bidirectional version of Dijkstra's algorihtm. Namely, we run two copies of Dijkstra's algorithm simultaneously: the first copy proceeds forward from the source, while the second copy proceeds backward from the destination. In our fire analogy, this would be like lighting up at the same time the two endpoints of the shortest path: as soon as the two fire fronts meet, the shortest path has been computed and thus the computation can be stopped. The result of one such experiment in the North American road network around Seattle is shown in Fig. 4.8b. Although the area explored by this bidirectional version of Dijkstra's algorithm is much smaller than the area explored by the classic Dijkstra's algorithm

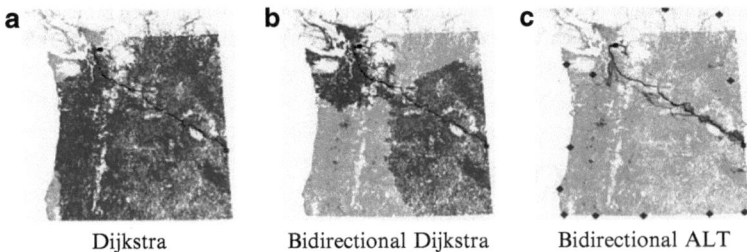

Dijkstra   Bidirectional Dijkstra   Bidirectional ALT

Fig. 4.8 Portion of the North American road network around Seattle explored by three different shortest path algorithms. The network contains about 1.6 million vertices and 3.8 million edges, and edge costs are average travel times. The *dark gray area* contains the vertices explored by each algorithm, while the shortest path is shown in *black*. (**a**) Dijkstra. (**b**) Bidirectional Dijkstra. (**c**) Bidirectional ALT

(in Fig. 4.8a), the bidirectional Dijkstra's algorithm still explores about 100,000 vertices.

The increasing popularity of navigation systems and the availability of online services for maps and driving directions which have to support millions of user requests per second, such as Google Maps or Bing Maps, have sparked in the last decade new research on faster single-pair shortest path algorithms. In November 2004, the first author of this chapter was visiting the Microsoft Research Labs in Silicon Valley. He was working with Andrew V. Goldberg on graph optimization problems, and they conceived the idea of launching a new challenge to the algorithmic community, where research teams from all over the world would have to compete to design and implement the fastest shortest path algorithm. David S. Johnson, Head of the Algorithms and Optimization Department at AT&T Research Labs in New Jersey, liked the idea. In 1990 David Johnson had started the DIMACS Implementation Challenges, a series of events devoted to realistic algorithm performance and to leading-edge design and implementations of algorithms. The Implementation Challenges, sponsored by the Center for Discrete Mathematics and Theoretical Computer Science (DIMACS), proved to be very influential in many areas of combinatorial optimization.

In September 2005, the *9th DIMACS Implementation Challenge*, devoted to the shortest path problem, was officially started and many leading research groups contributed to the event. After 1 year of hard work, the five finalists presented their algorithms and implementations in the final workshop, which took place in November 2006 at the DIMACS Center in Rutgers University, New Jersey. The final competition consisted of finding shortest paths for a sequence of 1,000 random cities in the US road network, which contains roughly 24 million vertices and 60 million edges. Each challenger had 24 h to look at the road graph and perform any required preprocessing on that input graph. The results of the challenge were surprising: on a low-end PC, all five finalists were able to answer a shortest path query in less than 3 ms on average. The fastest program, designed by Peter Sanders and Dominik Schultes [97], was able to answer a shortest path query in about 20 μs on average:

this is roughly 200,000 times faster than running Dijkstra's algorithm on the same computer!

The secret of the efficiency of those algorithms lies in their ability to look for the solution by exploring only a limited search space, without having to wander through unpromising areas of the entire input graph. To illustrate this concept, consider Fig. 4.8c, which shows the footprint left by the ALT algorithm by Goldberg and Harrelson [50] while looking for the same shortest path as the two algorithms of Fig. 4.8a, b. As can be easily seen from Fig. 4.8, the ALT algorithm explores only a tiny fraction of the vertices, centered around the shortest path to be searched for. In order to guide the search towards the final destination, Goldberg and Harrelson adapted A* search [56], a famous algorithm used for many problems in Artificial Intelligence. The details of their approach are spelled out in the next section.

4.9.1 The ALT Algorithm by Goldberg and Harrelson

Let s and t be, respectively, the source and destination vertex of an instance of the single-pair shortest path problem. The ALT algorithm by Goldberg and Harrelson [50] resembles Dijkstra's algorithm, with one notable exception: vertices are not considered according to their tentative distance computed from the source vertex s, but rather according to their combined tentative distance both from the source s and to the destination t. This requires it to maintain, for each vertex v, estimates of its distance from the source s and of its distance to the destination t. Note that the tentative distance $D(v)$ from the source s to each vertex v is already maintained (as an upper bound) by Dijkstra's algorithm, so we need only to worry about the tentative distance from each vertex v to the destination t. Similarly to the definition of the tentative distance $D(v)$, we do not know a priori the distance from v to the destination t. Thus, in the algorithm we make use of an estimate of this distance, possibly a lower bound of its exact value, denoted by $H(v)$. The next vertex that will be selected by the ALT algorithm is the vertex v in the set B which minimizes the quantity $D(v) + H(v)$.

The crux of the algorithm is to find good distance estimates $D(v)$ and $H(v)$ for each vertex v: the more accurate these distance estimates, the faster the algorithm, as it will explore only vertices centered around the shortest path. Indeed, in the special (and lucky) case where all our tentative distances coincide with exact distances, then it is not difficult to see that the algorithm will limit its search only to the set of shortest path vertices. So the natural question is to find good estimates $H(v)$ for the distance from vertex v to the destination t. To accomplish this task, Goldberg and Harrelson used a basic property of metric spaces, called *triangle inequality*: given any three vertices ℓ, t and v in a graph, the distance from ℓ to t does not exceed the sum of the distances from ℓ to v and from v to t (see Fig. 4.9). More formally, we have the following inequality:

$$d(\ell, t) \leq d(\ell, v) + d(v, t).$$

Fig. 4.9 Triangle inequality in a graph

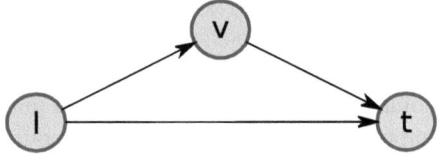

Note that we can rewrite the triangle inequality as

$$d(\ell, t) - d(\ell, v) \leq d(v, t).$$

Let t be the destination vertex for our shortest path, let v be a vertex to be extracted from the set B of the original Dijkstra's algorithm and let ℓ be any other vertex in the graph. If the distances from ℓ to t and from ℓ to v are known, the above inequality states that the quantity $d(\ell, t) - d(\ell, v)$ is a lower bound on the distance from v to t, and thus it can be used to compute $H(v)$ in the algorithm:

$$H(v) = d(\ell, t) - d(\ell, v).$$

The accuracy of this estimate for $H(v)$ depends on the particular choice of the vertex ℓ. Indeed if ℓ is closer to the destination t than to the vertex v, then $d(\ell, t) \leq d(\ell, v)$ and consequently $H(v) = d(\ell, t) - d(\ell, v) \leq 0$, which is not a very useful estimate. In order to be resilient to possibly bad choices of the vertex ℓ, Goldberg and Harrelson did not use only one vertex ℓ in the triangulation, but rather defined a set of vertices L, which they called *landmarks*, and considered several possible triangulations with respect to the vertices in L. To obtain a good estimate of $H(v)$, they considered the maximum value obtained over all possible choices of the landmark ℓ in L, as follows:

$$H(v) = \max_{\ell \in L} \{d(\ell, t) - d(\ell, v)\} \tag{4.1}$$

Now, it should be clear why the algorithm was called ALT, since it is based on A* search, Landmarks and Triangle inequalities. Note that a proper selection of landmarks is critical for the performance of the ALT algorithm. In particular, it follows from Eq. (4.1) that the initial choice of landmarks has an immediate impact on the quality and efficiency of the algorithm. For this reason, Goldberg and Harrelson designed and engineered several heuristics for selecting landmarks and evaluated them empirically. The landmarks that the ALT algorithm selected for the North American road network around Seattle appear as diamond dots in Fig. 4.8c: note that most of them are close to the edge of the road map. Once the landmarks are chosen, to apply Eq. (4.1) we still need to know all distances from every landmark to every vertex, which can be computed in an initial preprocessing phase. With all this information precomputed and stored in memory, the algorithm is now able to answer efficiently any single-pair shortest path query. The implementation used in the final competition of the DIMACS Implementation Challenge on shortest paths was a bidirectional variant of the ALT algorithm, where two searches proceed simultaneously (until they meet) from the two endpoints of the shortest path.

We observe that those recent algorithmic results, even though inherently bound to the "Reign of Quantity", using a metaphor loved by the French philosopher René Guénon (1886–1951), represent the apex of a research path that started one morning in 1956 in front of a coffee cup in Amsterdam. Although in recent decades research has produced major technological breakthroughs and has been of significant impact in everyday life, today's faster shortest path algorithms are still descendants of that famous common ancestor designed for the ARMAC demonstration, and preserve its fundamental spirit and basic properties. On the other hand, that common ancestor is in turn a descendant of an ancient thought, aimed at mimicking, and possibly understanding, the deepest mechanisms underlying Nature and Nature's ubiquitous ability to find the shortest route.

4.10 Bibliographic Notes

The shortest path problem arises in many application areas, particularly in communication and transportation networks. For this reason, it has been studied in many disciplines, including computer science, operations research and discrete mathematics. The original algorithm by Dijkstra is described in [30], while an excellent survey on algorithms for shortest paths is contained in the textbook by Ahuja et al. [4]. A basic ingredient for shortest path algorithms is the heap data structure; a general treatment of heaps can be found in a classic algorithmic textbook, such as the book by Cormen et al. [20]. The interested reader is referred to [110] for the low-level details of the classic binary tree heaps by Williams, and to [45] for a complete description of the Fibonacci heaps by Fredman and Tarjan.

Given their importance in many applications, shortest path algorithms have been investigated not only from the theoretical but also from the experimental viewpoint. In particular, a thorough experimental study of the practical performance of shortest path algorithms was carried out by Cherkassky et al. [18]. Other empirical studies on shortest paths were performed by Gallo and Pallottino [46], and by Divoky and Hung [34]. Many other implementations and experimental evaluations of shortest path algorithms were recently presented during the 9th DIMACS Implementation challenge [26]. Among the new algorithms proposed in this Challenge, we mention the algorithm by Sanders and Schultes [97] and the ALT algorithm by Goldberg and Harrelson [50], which is based on A*, a famous search algorithm proposed by Hart et al. [56].

Acknowledgements We thank Andrew V. Goldberg, Chris Harrelson, Haim Kaplan and Renato F. Werneck for making available to us the images in Fig. 4.8.

Chapter 5
Web Search

Paolo Ferragina and Rossano Venturini

Abstract Faced with the massive amount of information on the Web, which includes not only texts but nowadays any kind of file (audio, video, images, etc.), Web users tend to lose their way when browsing the Web, falling into what psychologists call "getting lost in hyperspace". Search engines alleviate this by presenting the most relevant pages that better match the user's information needs. Collecting a large part of the pages in the Web, extrapolating a user information need expressed by means of often ambiguous queries, establishing the importance of Web pages and their relevance for a query, are just a few examples of the difficult problems that search engines address every day to achieve their ambitious goal. In this chapter, we introduce the concepts and the algorithms that lie at the core of modern search engines by providing running examples that simplify understanding, and we comment on some recent and powerful tools and functionalities that should increase the ability of users to match in the Web their information needs.

5.1 The Prologue

Just 10 years ago, major search engines were indexing about one billion Web pages; this number has today exploded to about one trillion as reported in Google's blog by Alpert et al. [5]. Such growth is proportional to three orders of magnitude, thus leading everyone to talk about the *exponential* growth of the Web. But this number denotes only the amount of pages that are *indexed* by search engines and thus are available to users via their Web searches; the *real* number of Web pages is much larger, and in some sense unbounded, as many researchers observed in the past. This is due to the existence of pages which are dynamic, and thus are generated on-the-fly when users request them, or pages which are hidden in private archives, and thus can be accessed only through proper credentials (the so-called *deep Web*). At the extreme, we could argue that the number of (dynamic) pages in the Web is infinite, just take sites generating calendars.

Faced with this massive amount of information, which includes not only text but nowadays any kind of file (audio, video, images, etc.), Web users tend to lose their way when browsing the Web, falling into what psychologists call "getting lost in hyperspace". In order to avoid this critical issue, computer scientists designed in the recent past some sophisticated software systems, called *search engines*, that allow users to specify some keywords and then retrieve in a few milliseconds the collection of pages containing them. The impressive feature of these systems is that the retrieved pages could be located in Web servers spread all around the world, possibly unreachable even by expert users that have clear ideas of their information needs. In this chapter we will review the historical evolution and the main algorithmic features of search engines. We will describe some of the algorithms they hinge on, with the goal of providing the basic principles and the difficulties that algorithm designers and software engineers found in their development. This will offer a picture of the complexity of those systems that are considered as the most complex tools that humans have ever built. A commented literature concludes the chapter by providing pointers to several fundamental and accessible publications that can help readers to satisfy their curiosity and understanding of search engines and, more specifically, the wide and challenging research field known as Information Retrieval.

The following sections will cover all those issues, starting with an analysis of two networks that are distinct, live in symbiosis, and are the ground where search engines work: the Internet and the Web.

5.2 Internet and Web Graphs

As we saw in Chap. 2, a *graph* is a mathematical object formed by a set of *nodes* and a set of *edges* which represent relationships between pairs of nodes. Founded in the eighteenth century, Graph Theory has gained great importance in the last century as a means to represent entities in relation to each other. Nowadays, Graph Theory has further increased its importance being essential for the study of the Internet and its applications (such as the Web).

A network of computers (hereinafter simply referred to as a *network*) is a set of devices that send messages through electronic connections by means of cables, fiber optics, radio or infrared links. In our abstraction of a network, connections are represented with the edges of a graph whose nodes represent the devices. The Internet is actually a network of networks: institutions may own many computers which are connected to each other, but each institution enters the Internet as a single unit. These units are called *autonomous systems* (ASs) and constitute the nodes of the Internet graph. Each of these ASs could be either a user owning a single computer, or a set of computers connected together within the same building, or even very complex networks whose computers may be geographically far from

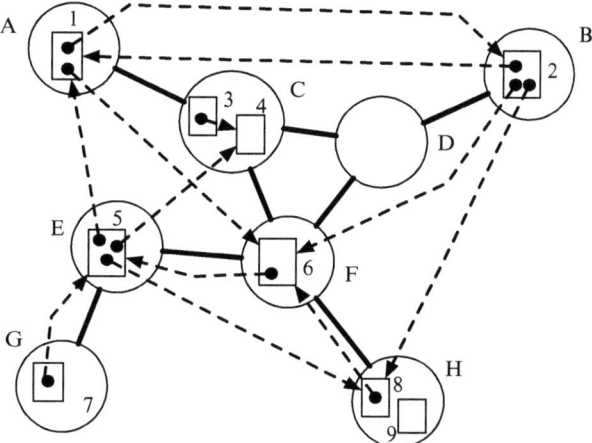

Fig. 5.1 The structure of the Internet and the Web graphs. *Circles* represent the autonomous systems, and edges marked with *solid lines* represent the connections between them. *Rectangles* represent Web pages, and edges marked with *dashed arrows* represent the links between them

each other. *Internet Service Providers* (ISPs) are examples of the latter typology of ASs. ISPs are companies which regulate the traffic of messages on the Internet by selling their services to other users.

Figure 5.1 shows a possible fragment of the Internet graph: nodes are drawn as circles connected by edges which are represented by continuous lines. Notice that edges can be traversed in both directions. A direct message from G to B, e.g., an e-mail, could follow the path $G - E - F - D - B$. User G is the sender who pays her service provider E, which regulates its costs with subsequent nodes. However, the real scenario is much more complicated. The Internet is huge: even if we do not know how many computers are connected (this question is indeed misplaced because the network topology is constantly changing), we can estimate that this number exceeds one billion, considering all the computers within the ASs. The path followed by a message is not determined beforehand and may even change during the transmission. This is a consequence of the enormous size and *anarchistic* structure of the Internet, which grows and evolves without any centralized control, and the inevitable continuous changes in the connections deriving from technical or maintenance issues. This behavior has characterized the network since its birth in the late 1960s and distinguishes the Internet from both telephone and electrical networks.

Traditional telephone networks work by using a methodology called *circuit switching*: two people talking on the phone are using the "channel" that has been reserved for their communication. The two network nodes establish a dedicated communication channel through the network before they start communicating. This channel works as if the nodes were physically connected with an electrical circuit. However, if something goes wrong, the communication is interrupted. The initial idea for the Internet was to resort to a mechanism called *message switching*: the routing of messages in the network is established node by node depending on the location of the recipient and the current level of traffic. At each step, the current node is responsible for choosing the next node in the path. As a side effect of this mechanism, a message sent to a close user may be sent through a path longer than necessary. This method was soon replaced by a close-relative method called *packet switching*, which is currently still in use. A (binary) message is divided into *packets* of a fixed length; this length is reported at the beginning of each packet together with its destination address. Sometimes it happens that packets of the same message are routed thought different paths and reach the destination in a order that differs from the original one. The recipient is then responsible for reordering the packets to (re-)obtain the original message. This mechanism guarantees that a message is always accepted by the network, even if it may take time before all of its packets reach their final destination and the message can be reconstructed. In a telephone network, instead, a phone number can be "busy" even if the recipient's phone is free, due to the saturation of the chain of connections which link the two locations involved in the call because of other calls.[1]

Now let us see what we mean by Web (or *www*, which is the acronym for the World Wide Web). Born in 1989 at CERN in Geneva and based on the known concept of *hypertext*, the Web is a set of documents called *pages* that refer to each other to refine or improve the same subject, or to draw a new subject in some relation to the first one. The luck of the Web is inextricably linked to the Internet. Pages of the Web are stored in the memories of computers on the Internet network, so that users can freely consult this collection of pages by moving from one page to another one via (hyper-)links, quickly and without leaving their PCs, regardless of the location of the requested pages. Seen in this way, the Web is represented as a graph whose nodes are pages and whose edges are the (hyper-)links between pairs of pages. Edges in this graph are *directed*, meaning that each edge can be traversed only in one predetermined direction. Figure 5.1 shows a portion of the Web graph in which the orientation of an edge is specified by means of an arrow.

We should immediately establish some properties of this graph being very important for search engines which collect, inspect and make Web pages available to users. First, the graph is literally huge, and its topology varies continuously due to the continuous creation or deletion of pages. Regarding the size of the graph, sometimes we read unfounded and fanciful stories: it is reasonably sure that search engines provide access to roughly hundreds of billions of pages. Nonetheless, the number of existing pages is larger, though many of them may be not directly available as we commented at the beginning of this chapter.[2] Another important observation is that, although the Web pages are stored in computers of the Internet,

---

[1]This phenomenon happened frequently years ago. Nowadays, it is rare due to the improvement of transmission techniques. However, it can still happen if we call foreign countries or if we try to make national calls in particular periods of the year (e.g., New Year's Eve).

[2]We usually refer with *indexable Web* to the set of pages that could be reached by search engines. The other part of the Web, called the *deep Web*, includes a larger amount of information contained in pages not indexed by search engines, or organized into local databases, or obtainable using special software. Nowadays an important area of research studies the possibility of extending the functionalities of search engines to the information stored in the deep Web. In this direction can be classified the initiative *open data*, see, for example, linkeddata.org.

```
      A B C D E F G H           1 2 3 4 5 6 7 8 9           1 2 3 4 5 6 7 8 9
  A   0 0 1 0 0 0 0 0       1   0 1 0 0 0 1 0 0 0       1   1 0 0 0 1 1 0 1 0
  B   0 0 0 1 0 0 0 0       2   1 0 0 0 0 1 0 1 0       2   0 1 0 0 1 2 0 0 0
  C   1 0 0 1 0 1 0 0       3   0 0 0 1 0 0 0 0 0       3   0 0 0 0 0 0 0 0 0
  D   0 1 1 0 0 1 0 0       4   0 0 0 0 0 0 0 0 0       4   0 0 0 0 0 0 0 0 0
  E   0 0 0 0 0 1 1 0       5   1 0 0 1 0 0 0 1 0       5   0 1 0 0 0 2 0 0 0
  F   0 0 1 1 1 0 0 1       6   0 0 0 0 1 0 0 0 0       6   1 0 0 1 0 0 0 1 0
  G   0 0 0 0 1 0 0 0       7   0 0 0 0 1 0 0 0 0       7   1 0 0 1 0 0 0 1 0
  H   0 0 0 0 0 1 0 0       8   0 0 0 0 0 1 0 0 0       8   0 0 0 0 1 0 0 0 0
                             9   0 0 0 0 0 0 0 0 0       9   0 0 0 0 0 0 0 0 0

             I                            W                            W²
```

Fig. 5.2 Adjacency matrices I and W for the Internet and the Web graphs of Fig. 5.1, and the square matrix W^2

these two graphs do not have any other relationship between each other. Referring to the example of Fig. 5.1, page 1 has a link to page 2 but there is no direct link between the two nodes A and B of the Internet (i.e., the computers containing these pages). On the other hand, the two nodes C and F are connected but there are no links between the pages contained therein. Another difference between these two graphs is that the edges of the Web are directed, while those of the Internet are not. Moreover, the graph of the Internet is *strongly connected*, i.e., there always exists a path that connects any two nodes, unless a temporary interruption of some connection does occur. This, however, does not happen in the Web graph where there are nodes that cannot be reached by others. This may happen for several reasons: a node cannot be reached by any other node since it has only outgoing links (nodes 3 and 7 in the figure), a node cannot reach other nodes since it has only incoming links (node 4), or a node is completely disconnected since it has no links at all (node 9). In the box below we present a deeper algorithmic elaboration about these important characteristics.

Adjacency Matrix and Paths of a Graph

A graph G with n nodes can be represented in a computer through an *adjacency matrix* M having n rows and n columns. Rows and columns are in correspondence with nodes of G. The cell of M at row i and column j, denoted with $M[i, j]$, corresponds to the pair of nodes i and j. We set $M[i, j]$ equal to 1 whenever G has an edge from node i to node j; $M[i, j]$ is 0 otherwise. The Internet graph and the Web graph of Fig. 5.1 have, respectively, the adjacency matrices I of size 8×8 and W of size 9×9. These matrices are shown in Fig. 5.2.

Notice that any undirected graph induces an adjacency matrix which is symmetric with respect to its main diagonal, because if the edge (i, j) exists then so does the edge (j, i), hence $M[i, j] = M[j, i] = 1$. We observe that the Internet graph is undirected and, thus, matrix I is symmetric.

(continued)

(continued)

For an example, edge $A - C$ can be traversed in both directions and, thus, $I[A, C] = I[C, A] = 1$. In directed graphs, as the Web graph, the matrix may be asymmetric and so there exist entries such that $M[i, j] \neq M[j, i]$. In our example we have $W[1, 2] = 1$ and $W[2, 1] = 1$ because there are two distinct edges going in both directions. However, we also have cases in which only one edge between two nodes is present (e.g., we have $W[1, 6] = 1$ and $W[6, 1] = 0$).

In mathematics the square of an $n \times n$ matrix M is an $n \times n$ matrix M^2 whose cells $M^2[i, j]$ are computed by a way different than the standard product of numbers. Each cell $M^2[i, j]$ is, indeed, obtained by multiplying the ith row and the jth column of M according to the following formula:

$$M^2[i, j] = M[i, 1] \times M[1, j] + M[i, 2] \times M[2, j] + \ldots + M[i, n] \\ \times M[n, j]. \tag{5.1}$$

This formula has a deep meaning that we will illustrate through our example matrix W of Fig. 5.2. Take $W^2[6, 4]$, which is equal to 1 because the sixth row and the fourth column of W are, respectively, equal to [0 0 0 0 1 0 0 0 0] and [0 0 1 0 1 0 0 0 0]. So that the formula above returns the value 1 since there is a pair of multiplied entries ($W[6, 5]$ in row 6 and $W[5, 4]$ in column 4) that are equal to 1.

$$W^2[6, 4] = W[6, 1] \times W[1, 4] + W[6, 2] \times W[2, 4] + \ldots + W[6, 9] \\ \times W[9, 4] \\ = 0 \times 0 + 0 \times 0 + 0 \times 1 + 0 \times 0 + 1 \times 1 + 0 \times 0 + 0 \times 0 \\ + 0 \times 0 + 0 \times 0 \\ = 1$$

Interestingly, there is a deeper reason behind the value 1 obtained by multiplying $W[6, 5]$ and $W[5, 4]$. Since these two cells indicate that in the Web graph there is one edge from node 6 to node 5 and one edge from node 5 to node 4, we conclude that there exists a path that goes from node 6 to node 4 traversing exactly two edges. Therefore, each cell $W^2[i, j]$ indicates the number of distinct paths from node i to node j that traverse exactly two edges. We can understand this statement better by further examining other entries of the matrix W^2. We have $W[1, 2] = 1$ but $W^2[1, 2] = 0$ because there is an edge from node 1 to node 2 but there does not exist any path of length 2 connecting these two nodes (see Fig. 5.1). Furthermore, we have

(continued)

(continued)

$W^2[2, 6] = 2$ because the second row of W is [1 0 0 0 0 1 0 1 0] and its sixth column is [1 1 0 0 0 0 0 1 0]:

$$\begin{aligned} W^2[2,6] &= W[2,1] \times W[1,6] + W[2,2] \times W[2,6] + \ldots + W[2,9] \\ &\quad \times W[9,6] \\ &= 1 \times 1 + 0 \times 1 + 0 \times 0 + 0 \times 0 + 0 \times 0 + 1 \times 0 + 0 \times 0 \\ &\quad + 1 \times 1 + 0 \times 0 \\ &= 2 \end{aligned}$$

Here we find two pairs of 1s: $W[2, 1] = W[1, 6] = 1$ and $W[2, 8] = W[8, 6] = 1$, meaning that the Web graph has two paths of length 2 connecting node 2 to node 6 (i.e., $2 - 1 - 6$ and $2 - 8 - 6$).

Following the same rule we can proceed in the calculation of successive powers W^3, W^4, ... of the matrix W. The entries of these matrices indicate the number of paths of length 3, 4, ... between pairs of nodes, and thus the number of links that we must follow in order to move from one Web page to another. For example, the elements of W^3 are obtained by resorting to Eq. (5.1), as the product of a row of W^2 and a column of W (or vice versa). We have thus:

$$W^3[7, 6] = 1 \times 1 + 0 \times 1 + 0 \times 0 + 1 \times 0 + 0 \times 0 + 0 \times 0 + 0 \times 0 + 1 \times 1 + 0 \times 0 = 2$$

which corresponds to the two paths of length 3 from node 7 to node 6 (i.e., $7 - 5 - 1 - 6$ and $7 - 5 - 8 - 6$).

Classical algorithms compute a power of a matrix by taking two powers with smaller exponents and by applying Eq. (5.1) to one row and one column of each of them. Since the number of these pairs is n^2, and since the calculation of Eq. (5.1) requires a time that is proportional to n (in fact, the expression contains n multiplications and $n - 1$ additions), the total computation time required by one matrix multiplication is proportional to (i.e., *is order of*) n^3. Of course, the actual time required by an algorithm also depends on the computer and on the programming language in use. n any case, a *cubic* behavior like this is undesirable: it means, for example, that if the number of nodes of the graph doubles from n to $2n$, the time grows from n^3 to $(2n)^3 = 8n^3$, i.e., it becomes eight times larger. As we shall see in a following section, computing powers of matrices is one of the main ingredients to establish an order of importance (*ranking*) among Web pages. However, considering that the Web graph has hundreds of billions of nodes, it is unfeasible to perform

(continued)

> **(continued)**
>
> this computation on a single computer. The solution usually adopted consists of dividing the work among many computers by resorting to techniques of *distributed computation*, which are, however, much too complex to be discussed here.

5.3 Browsers and a Difficult Problem

A *Web site* is a group of related Web pages whose content may include text, video, music, audio, images, and so on. A Web page is a document, typically written in plain text, that, by using a special language, specifies the components of that page (e.g., text and multimedia content) and the way in which they have to be combined, displayed and manipulated on a computer screen. A Web site is hosted on at least one *Web server* and can be reached by specifying the address of one of its pages. Each page has, indeed, its own numeric address which is, for convenience, associated to a *textual name* which is easier to remember (called URL). For example, www.unipi.it/index.htm is the name of the main Web page of the University of Pisa, while its numeric address of the hosting Web server is `131.114.77.238`. A Web site is characterized by its domain name, e.g., `unipi.it`, and by its main page, which is the starting point to visit the other secondary pages of the site. The URLs of these pages are refinements of the domain name through a hierarchical structure expressed by means of a path. For example, www.unipi.it/research/dottorati is the address of the page that contains the list of PhD courses of the University of Pisa.

A *browser* is a software that allows Web users to visualize a page on the screen of a computer by specifying its URL. After the first Web browser developed at CERN for the original Web, many commercial products were developed, e.g., Netscape Navigator, Internet Explorer, Firefox, Chrome, and many others. In order to speed up the access to Web pages by browsers and other software/applications, network engineers have designed special nodes that collect large groups of page copies. These nodes are called *caches*, and *caching* refers to the activity of storing pages in them. Figure 5.3 shows a typical organization of connections and caches in the Internet graph.

A key role in the network is played by *proxies*, which are computers serving as a local cache within an AS. These computers keep copies of the most frequently requested Web pages by users of the AS. These pages are typically news sites, popular social networks, search engines, and so on. The cached pages are either from outside the AS, or from inside the AS. In the example of Fig. 5.3, users A and B reside inside the same AS, so if they have recently requested the same page, then this page is probably stored in proxy 1 of their AS. In addition, proxies cache Web pages which are stored in the AS and are requested frequently from computers

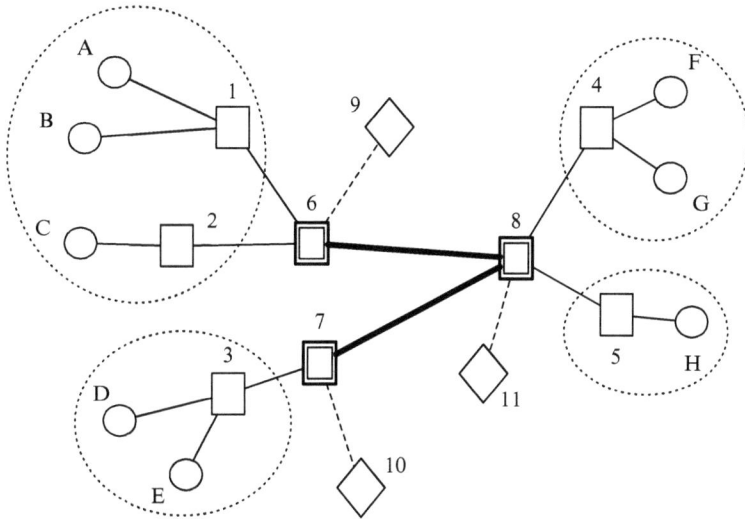

Fig. 5.3 Several levels of caching for Web pages are shown. *Circles* labeled with $A, B, \ldots, H$ are computers with a browser which store in their memory the most recently accessed Web pages. Regions enclosed with *dotted lines* are ASs, while *rectangles* represent their corresponding proxies. *Double rectangles* represent routers, which are devices that route the messages flowing through the network. Finally, *diamonds* represent large CDN subnetworks

outside that AS. In this way, these frequent requests are served immediately without the need of traversing the whole AS. In the example of Fig. 5.3, if pages stored in the computers A and B are requested frequently from other users of the network, copies of them may be kept in proxy 1, which is responsible for answering requests coming from outside. Caching can be structured in a hierarchical way, by introducing the so-called *Content Delivery Networks* (CDN), which are subnetworks of computers that provide caching at geographic level. It goes without saying that the same Web page can be replicated in many proxies and CDNs, provided that this improves its delivery to the requesting browsers and Web applications.

An interesting problem is that of distributing Web pages in the caches of the network with the aim of minimizing the overall amount of time required to access the Web pages that are more likely to be requested by users in a certain period of time. As we will see, from an algorithmic point of view, this problem is practically insolvable because all known solutions have exponential time complexity. In fact, this problem belongs to a class of hard problems for which we can compute efficiently only approximated solutions.[3] We will support this claim by introducing the well-known Knapsack Problem, discussed in Sect. 2.3.2. What is nice in this

---

[3]This class of problems, called *NP-hard* problems, is extensively discussed in Chap. 3.

Object	1	2	3	4	5	6	7
Weight	15	45	11	21	8	33	16
Value	13	25	14	15	6	20	13
	1	0	1	1	0	0	1

Fig. 5.4 An example of the Knapsack Problem: we have to choose a subset of objects that maximizes their total value and satisfies a constraint on their total weight, which should be at most $C = 70$. The last row represents the optimal subset $\{1, 3, 4, 7\}$, which is encoded by indicating with a 1 an object included in the subset and with a 0 an object not included in it. The total weight is $15 + 11 + 21 + 16 = 63 < 70$, while the total value is $13 + 14 + 15 + 13 = 55$. This is optimal in that no other subset of weight smaller than 70 has a larger value

discussion is that, although these two problems appear very different, they turn out to be highly related in terms of the time required to compute their solution.

Let us consider Fig. 5.4 in which we have seven objects, numbered from 1 through 7, and a knapsack of maximum capacity $C = 70$. For each object we report its weight and its value (e.g., object 1 has weight 15 and value 13, object 2 has weight 45 and value 25, and so on). In the last row of the table we represent a subset of objects by using a bit that indicates whether the corresponding object has been selected (value 1) or not (value 0). We recall that the Knapsack Problem identifies the subset of objects which maximizes the total value, provided that its total weight is at most C. As already observed in Chap. 2, it is strongly believed that this problem does not admit any solution which is more efficient than the one that considers all possible subsets of objects and discards the ones having total weight larger than C. Since the number of possible subsets of n objects is 2^n (including the empty one), this solution requires exponential time and thus it is unfeasible even for a small number of objects.[4]

In the box below we will show the existing relation between the Knapsack Problem and the problem of distributing Web pages in a cache (Cache Problem). We observe that a more general formulation of the latter problem is more complicated than the one presented here. For example, in real applications it is typical to consider also the probability of accessing a certain page, the frequency of requests arriving from each AS, the time variation of these quantities, and so on. However, it suffices to show that the simplified problem is still hard in order to prove the hardness of any more general formulation. In our case, it can be proved that the Cache Problem is *at least as hard as* the Knapsack Problem. This proof is called *reduction* and, specifically, consists of showing that whenever there exists an algorithm that solves the Cache Problem in less than exponential time, the same algorithm can be applied with simple changes to the Knapsack Problem and solve it in less than exponential time too. Since this event is considered to be impossible,

---

[4]For example, for $n = 20$ objects we have $2^{20} > 1$ millions of subsets.

one can reasonably assume that such a "surprisingly efficient algorithm" for the Cache Problem does not exist; so this problem can be addressed only by means of exponential solutions or, efficiently, by returning approximate solutions. Space limitations and the scope of this chapter do not allow us to detail the reduction that links the Cache Problem to the Knapsack Problem; rather we content ourselves by showing a simpler result, namely, that an approximate solution for the Cache Problem can be derived from an approximate solution to the Knapsack Problem.

The Cache Problem
Let us assume that the network has k ASs, denoted with $AS_1, AS_2, \ldots, AS_k$, n pages, denoted with $p_1, p_2, \ldots, p_n$, and just one cache that stores copies of Web pages for a total size of at most B bytes. Furthermore we assume that each page is stored in its AS, that all pages are requested with the same probability, and that all ASs access the same number of pages. As in the Knapsack Problem, we define an array W of page weights so that $W[j]$ is the size in bytes of the file representing page p_j. It is a little bit harder to define an array of "values" for the Cache Problem that mimics the array of values in the Knapsack Problem. For this aim we use A^j to indicate the AS owner of page p_j; we use $d(i, j)$ to denote the distance, expressed in number of hops, that separate the generic AS_i from A^j; and we use $c(i, j)$ to indicate the cost that AS_i has to pay, in terms of number of hops, to obtain the page p_j. This cost may depend on the choice between placing or not placing the page p_j in the single cache we assumed to exist in the network. Indeed, if p_j is in the cache and the distance between AS_i and the cache is smaller than $d(i, j)$, we have $c(i, j) < d(i, j)$. In any other case, the value of $c(i, j)$ is equal to $d(i, j)$. We can then introduce the value $u(i, j) = d(i, j) - c(i, j)$, which expresses the gain for AS_i of having the page p_j copied in cache. At this point we are ready to define the value $V[j]$ of p_j as the total gain of placing p_j in the cache, summing over all ASs: $V[j] = \sum_{i=1\ldots k} u(i, j)$. Finally, the "reduction" to the Knapsack Problem can be concluded by taking an auxiliary array A to indicate the subset of pages to be cached.

At this point, we could solve the "synthetic" Knapsack Problem either exactly in exponential time (by enumerating all subsets), or approximately in polynomial time. In this second case, we could choose the objects (pages) in order of decreasing ratio value (gain) versus weight (byte size) until the capacity C of the knapsack (cache) is saturated. The behavior of this simple algorithm is shown in Fig. 5.5. However there do exist solutions to the Cache Problem that do not pass through this "algorithmic reduction": the simplest one consists of caching the most popular Web pages. Clearly, there do also exist more sophisticated approximation algorithms that exploit knowledge about the topology of the network and the frequency of distribution of the

(continued)

Object	3	1	7	5	4	6	2
Weight	11	15	16	8	21	33	45
Value	14	13	13	6	15	20	25
Value/Weight	1.27	0.87	0.81	0.75	0.71	0.61	0.55

Fig. 5.5 The objects of Fig. 5.4 are sorted by decreasing values of the ratio $V[i]/W[i]$. The heuristically chosen subset of objects is $\{3, 1, 7, 5\}$, its total weight is 50 and its total value is 46. Recall that the optimal subset has weight 63 and value 55, as shown in Fig. 5.4

(continued)

page requests. For example, the cache in a CDN C_i may give a larger priority to a page p_i having a high value of the product $\pi_j \times d(i, j)$, where π_j is the frequency of request for page p_j.

5.4 Search Engines

Browsers are fundamental tools for navigating the Web, but their effective use imposes that users clearly know their information needs and where the Web pages matching them are located in the Web. However, it is common that a user has only a partial knowledge of her information need, and wants to find pages through their content without necessarily knowing their URL. Search engines are designed to address this goal and are indispensable for finding information in the huge graph of available Web pages. Bing, Google and Yahoo! are the three most famous search engines available to Web users to match their information needs. They are based on similar algorithmic principles, which are nevertheless implemented differently enough to show, in response to the same user query, a different set of result pages. Here we will neither provide a comparison of different search engines, nor we will discuss how to implement effective user queries; rather we will limit ourselves to the more instructive description of the algorithmic structure of any modern search engine, detailing some of its components.

One of the key features of the search task is that it must be executed fast over a huge set of indexed pages. To this aim, each search engine resorts to a large number of computers grouped in different data-centers distributed around the world. Although many companies are reluctant to reveal precise information about their data-centers, it is estimated that each search engine deploys hundreds of thousands of computers organized into subnetworks, each of which provides different functions. We can distinguish these functions into two main categories (detailed in the following pages): those intended for building a huge index of the

Web pages, and those intended for resolving in the best and the fastest possible way the queries submitted by the users.

In the first category, we can identify several important algorithmic steps: the *crawling* of the Web, the analysis of the Web graph and the parsing of the crawled pages, and finally the construction of an *index* containing all relevant information to match efficiently the user queries. All these tasks are repeated at regular time intervals in order to keep the index (and therefore the results provided to the user queries) updated with respect to the continuous variations of the Web.

In the second category, we can also identify several important algorithmic steps which are executed at each user query and mainly consist of consulting the current index in order to discover the *relevant* pages for that query, ranking these pages in order of relevance, and possibly applying some classification or clustering tools that aim at offering different (and eventually more meaningful) views on the returned results. All these steps have as ultimate goal the one of satisfying in the best and the fastest way the information need hidden within the user queries.

A user query is typically formed by a sequence of *keyword*s. The process that leads the user to choose a specific set of keywords is critical since it significantly influences the quality of the results reported by the search engine. It is clear that an information need may be correctly settled by different groups of results. However, the actual relevance of these groups depends on the user submitting the query and her current information need, which may change, even drastically, from one user to another, and indeed it can change even for the same user depending on her specific interests at the time the query is issued. For example, the query **Lufthansa** for a user may have a *navigational* goal because she wants to find the homepage of the airline, a *transactional* goal because she might wish to buy a ticket, or an *informational* goal because she is interested in gathering some information regarding the company. And of course, the same user could issue queries at different times having different goals.

If the user is not completely satisfied by the results returned by the search engine, she could refine her query by adding keywords, or she could more clearly specify her intention by reformulating the query itself. However, this rarely happens: statistics show that more than 80 % of the queries are formed by only two keywords and their average number is around 2.5. Add to this that most users look at only the first page of results. This behavior is driven by not only user laziness in composing selective queries and browsing the returned results, but also in the intrinsic difficulty for users to model their information needs by means of appropriate keywords.

Despite all these problems, modern search engines perform their tasks very efficiently and provide very relevant results. Moreover, they are improving every day, thanks to intensive academic and industrial research in the field. In the following we will describe the salient operations performed by search engines noticing that, not surprisingly, many algorithms usually employed are not publicly known.

5.4.1 Crawling

In the slang of Internet, the term *crawling* refers to the (un-)focused retrieval of a collection of Web pages with the purpose of making them available for subsequent analysis, cataloguing and indexing of a search engine. A *crawler*, also named *spider* or *robot*, is an algorithm that automatically discovers and collects pages according to a properly-designed traversal of the Web graph. The reader should not confuse browsers with crawlers: the former retrieve and visualize specific pages indicated by a user via their URL; the latter retrieve and collect pages which are automatically identified via proper Web-graph visits. The following box details the algorithmic structure of a crawler.

> **Crawling Algorithm**
> A crawler makes use of two data structures called *queue* and *dictionary*, whose functionalities are close to the ones these terms assume in the context of transports (a queue of cars) and linguistics (a dictionary of terms), respectively. A queue Q is a list of elements waiting to be served. The element that is placed at the head of Q is the next that will be served and, when this happens, the element will be removed. In this way, the second element will become the new head of the queue. This operation is denoted by $Q \to e$ and indicates that element e is released outside. A new element is always inserted at the end of Q, and it will be served after all elements currently in Q. This operation is denoted by $e \to Q$.
>
> A dictionary D is a set of elements (not necessarily words in some natural language) waiting to be examined. In this case, however, there is more flexibility than in the queue regarding the operations that can be supported. What concerns us here is, indeed, that there is an efficient way to determine whether a particular element e is already in the dictionary D and possibly remove it (denoted by $D \to e$). The dictionary is built by means of insertions of new elements (denoted by $e \to D$). Notice that, since we wish to perform fast searches in the dictionary D, we have to carefully insert and organize the elements into it.
>
> Now we will introduce a (necessarily simplified) crawling algorithm whose fundamental steps are reported in Fig. 5.6. We first notice that the owner of a site could forbid the crawling of some Web pages, by adding a special file called `robots.txt`, which specifies which pages can be downloaded and which cannot.
>
> The crawling algorithm deploys a queue Q containing addresses of Web pages to be processed, two dictionaries D_{urls} and D_{pages} containing, respectively, the addresses of the Web pages already processed and an archive of

(continued)

(continued)

information extracted from those pages. Initially, both D_{urls} and D_{pages} are empty, while Q contains a set of addresses that are the starting seeds of the crawling process. Not surprisingly, the choice of these initial seed pages is fundamental to quickly reach the most relevant part of the Web. Seed pages are usually Web portals (e.g., DMOZ, Yahoo!), educational sites (e.g., Wikipedia and universities), news and social-network sites, since they contain pages that point to important and popular resources of the Web.

The algorithm of Fig. 5.6 is not trivial. When a new link U' is found in a crawled page, its address is inserted in Q ready to be processed in the future. This is done only if the link U' is not already present in D_{urls}, which means that its text $T(U')$ has not been downloaded yet. Notice that the same link U' may be contained in several pages which are discovered by the crawling algorithm before that U''s content is downloaded. Thus, this check ensures that U' will be downloaded and inserted in D_{pages} only once.

Obviously, state-of-the-art crawling algorithms are more complex than the one presented here, and have to include sophisticated functionalities and optimizations. One of the most important issues regards the fact that the Web changes at such high rate that, as estimated, we have a 30% renewal every year. So the crawlers must usually be trained to follow the more rapid variations (think, e.g., news sites), and designed to be as fast as possible in making "one scan of the Web" in order to keep the search engine index as fresh as possible. Moreover, the crawler should reduce the interactions with the crawled sites as mush as possible, in order to not congest them with continuous requests, and it should make use of advanced algorithmic techniques in *distributed computing* and *fault tolerance*, in order to ensure that it will never stop its operations. Therefore, the design and development of an efficient and effective crawler is much more complicated than what a reader could deduct from the algorithm reported in Fig. 5.6. The reader interested in those algorithmic details may look at the literature reported in Sect. 5.6.

We conclude this section by observing that the design of a crawler has to be optimized with respect to three parameters: the maximum number N of Web pages that can be managed before its algorithms and data structures are "overwhelmed" by the size of D_{url}; the speed S with which the crawler is able to process pages from the Web (nowadays crawlers reach peaks of thousands of pages per second); and, finally, the amount of computational resources (CPU, memory and disk space) used to complete its task. Clearly, the larger are N and S, the higher is the cost of maintaining the queue Q and the dictionaries D_{url} and D_{pages}. On the other hand, the more efficient is the management of Q, D_{url} and D_{pages}, the lower is the amount of computational resources used and the consumption of energy. The latter is nowadays an extremely important issue, given the high number of computers used to implement the modern search engines.

Crawling Algorithm

- Input: $\{u_1, \ldots, u_k\}$ an initial set of addresses of Web pages;
- Output: D_{urls} and D_{pages}.

1. Insert $u_1, \ldots, u_k$ into the queue Q;
2. Repeat until Q is non-empty
3.   Extract $Q \to u$ the next page-address u from Q;
4.   If $u \notin D_{\text{urls}}$, then
5.   Request the file `robots.txt` from the site of u;
6.   If this file allows to access page u, then
7.   Request the text $T(u)$ of the page u
8.   Insert $u \to D_{\text{urls}}$
9.   Insert $T(u) \to D_{\text{pages}}$
10.   Parse $T(u)$, and for any link u' in $T(u)$
11.   if $u' \notin D_{\text{pages}}$, add $u' \to Q$

Fig. 5.6 A simple crawler using the urls $\{u_1, \ldots, u_k\}$ as initial seed set

5.4.2 The Web Graph in More Detail

At this point it is natural to ask how large the Web graph is and what is the structure of its interconnections, since the effectiveness of the crawling process is highly dependent on these characteristics.

We have already noticed that the Web is huge and rapidly changing. There is, therefore, no hope that a crawler can collect in D_{pages} all the existing Web pages; so it has necessarily to be resigned to obtaining only a subset of the Web which, hopefully, is as broader and more relevant as possible. In order to optimize the quality of the collected pages, the crawler has to perform a visit of the graph which is inevitably more focused and complex than the one used in Fig. 5.6. For this purpose, the crawler uses a more sophisticated data structure, called *priority queue* (see Chap. 4), that replaces the simple queue Q and extracts its elements depending on a priority assigned to each of them. In our case the elements are Web pages and the priorities are values related to the relevance of those pages. The higher the priority, the sooner the crawler will process the page and download its neighbors. The objective is that of assigning low priority to pages with a lower relevance or that have been already seen, or to pages that are part of a site which is too large to be collected in its entirety. To model this last case, we take into account the *depth* of a page in its site as a measure of its importance. The depth is measured as the number of forward slashes in its URL address (for example, the page http://www.unipi.it/ricerca/index.htm is less deep than the page http://www.unipi.it/ricerca/dottorati/index.htm, and thus is assumed to be more important in that site and hence crawled first). Recent studies have shown that the depth and the number and quality of the links incoming and outgoing from a page are effective indicators for the assignment of these priorities.

5 Web Search

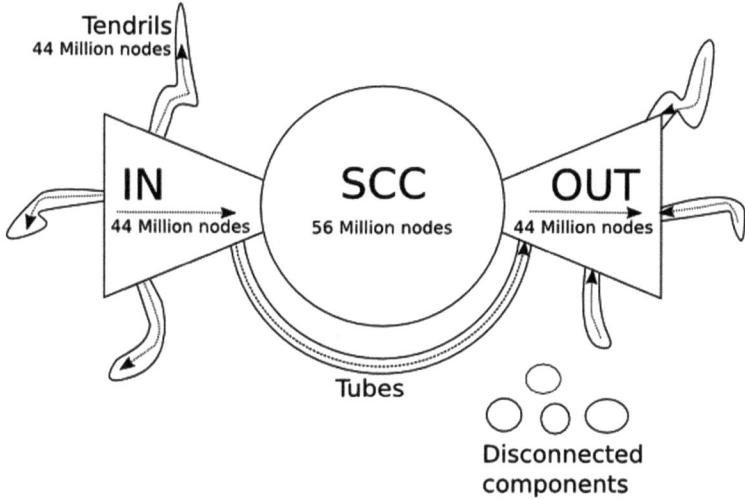

Fig. 5.7 The characteristic "bow" shape of the Web graph (1999). The subgraphs SCC, IN, OUT, tendrils and tubes, each consist of about one quarter of the nodes in the total graph

The structure of the graph significantly influences the behavior of the crawler. We consider two extreme examples that make this fact more evident. If the graph was made up of many disconnected components, the seed set of the crawler should contain at least one address for each of them; if the graph was instead made up of one (long) chain of pages, the seed set should contain the heading pages of this chain in order to guarantee that the graph visit traverses most of the Web. In November 1999 a study, now classic, analyzed the structure of a subgraph of the Web of that time formed by about 200 million pages. It turned out that this portion of the Web did not recall the two previous extreme examples, but it consisted of four main components shown in Fig. 5.7, all having about the same size: a strongly connected component, denoted SCC and called *core*,[5] a subgraph IN with paths that end up in pages of SCC, a subgraph OUT with paths that start at SCC, and a number of tendrils and tubes, namely pages linked in chains that do not pass through SCC or are completely isolated. These findings were later confirmed by studies carried out on larger and more recent samples of the Web: they actually showed not only that the sampled graph always has the form indicated in Fig. 5.7 and its components have about the same sizes, but also that the graph structure has some *fractal property* which leads any sufficiently large subgraph to have the same structure as its original containing graph. These surprising results are nowadays justified by mathematical studies on the laws that regulate the growth of the networks.

---

[5] Recall that a graph is strongly connected if and only if there exists a path that connects any pair of its nodes.

This structure of the Web graph suggests to insert in the seed set of the crawler pages chosen from IN or SCC. The problem is how to efficiently determine candidates from these two sets of nodes before a visit of the Web graph is performed. If candidates are chosen randomly from the Web, then we would have a probability 1/2 that each selected page is either in IN or in SCC, given that each of these sets is 1/4 of the total graph. Thus, it would be enough to choose a small number of candidates to be reasonably sure to start the crawling from IN or SCC. Unfortunately, this strategy is difficult to implement because there are neither available list of all the addresses of existing pages nor it is clear how to perform uniform sampling without this list. So the typical choice of existing crawlers is the simple one sketched above, and consists of taking as seeds the portals and any other sites that can lead to good pages of the Web and probably lie in IN or SCC.

5.4.3 Indexing and Searching

The pages collected by the crawler are subsequently processed by an impressive number of algorithms that extract from them a varied set of information that is stored in proper data structures, called *index*es, allowing it to efficiently answer the user queries. Entire subnets of computers are dedicated to this significant task in order to perform it in a reasonable amount of time.

Hereafter we will talk about *document*s, instead of pages, and with this term we will refer to the content of a Web page p plus some other information about p itself collected during the crawling process. An example of such additional information is the so-called *anchor text*, which corresponds to a portion of text surrounding a link to p in another page. Given that page p may have many incoming links, page p may have many anchor texts written by authors who are possibly different from the author of p. An anchor text may be therefore a particularly reliable and important piece of information because, presumably, it describes succinctly the content of p. Not surprisingly, search engines give great importance to anchor texts and use them to extend the content of Web pages, because they typically use a different set of words to describe their content and thus allow search engines to extend the results of a query.

For example, let us assume that a page p contains pictures of various species of insects but does not contain the word "insect(s)" in its body, or even does not contain any text at all, consisting just of that picture. Nevertheless it is possible that a passionate entomologist wrote his own Web page with a link to page p and annotated this link with the phrase "beautiful images of insects". This piece of text is an anchor for p, so the words in "beautiful images of insects" are added to those found in p and are considered highly characteristic for this page. Therefore a query "insects" would find p, even if it does not contain that word.

Unfortunately, as often happens in the Web, a valuable use of information is followed by a malicious one of it. In 1999 the first result reported by Google for the query "more evil than Satan" was the homepage of Microsoft. This phenomenon was eventually the result of the creation of many pages containing links to the homepage of Microsoft with anchor text "more evil than Satan". This awkward situation was resolved by Google in a few days, but a similar accident happened again in November 2003 with the query "miserable failure" and the page returned by Google as the first result was the homepage of the former U.S. President George W. Bush. This type of attack is nowadays called *Google bombing*, and it was repeated many other times in languages other than English.

Once the search engine has crawled a huge set of documents, it analyzes them to extract the *terms* contained therein and inserts these terms in a dictionary. Checking the presence of query terms in this dictionary is the first step performed during a query resolution. We observe that a query term may be not just a word, but any sequence of alphanumeric characters and punctuations because it may represent telephone numbers (911 or 1-800-237-0027), abbreviations (e-ticket), models of our favorite devices (N95, B52, Z4), codes of university courses (AA006), and so on. We cannot go into the details of efficient implementations of dictionary data structures, but we observe here that it is unfeasible to implement keyword searches through a linear scan of the dictionary, because this would take too much time.[6] It is thus crucial to have appropriate algorithms and data structures to manage efficiently, both in time and space, these numerous and long sequences of characters. One approach could be the dichotomous search algorithm presented in Fig. 2.24; more efficient solutions are known, mainly based on *tries* and *hash tables*, so we refer the reader to the literature mentioned at the end of this chapter.

The dictionary is just a part of the mass of information extracted by a search engine from the crawled documents, during the so-called *indexing* phase. The overall result of this phase is the construction of a data structure, called an *inverted list*, which is the backbone of the algorithms answering the user queries. An inverted list is formed by three main parts: the above dictionary of terms, one list of occurrences per term (called *posting list*), plus some additional information indicating the importance of each of these occurrences (to be deployed in the subsequent ranking phase). The word "inverted" refers to the fact that term occurrences are not sorted according to their position in the text, but according to the alphabetic ordering of the terms to which they refer. So inverted lists remind the classic *glossary* present at the end of a book, here extended to represent occurrences of all terms present into a collection of documents.

The posting lists are stored concatenated in a single big array kept in memory. The URLs of the indexed documents are placed in another table and are

---

[6]Several experimental results have shown that the number n of distinct terms in a text T follows a mathematical law that has the form $n = k|T|^\alpha$, with k equal to a few tens, $|T|$ being the number of words of the text, and α approximately equal to $1/2$. The actual size of the Web indexed by search engines is hundreds of billions of pages, each with at least a few thousand terms, from which we derive $n > 10 \times 10^6 = 10^7$. Thus, the dictionary can contain hundreds of millions of distinct terms, each having an arbitrary length.

Fig. 5.8 Indexing by using inverted lists. From the dictionary D we know that the list of documents containing the term "football" starts at position 90 in the array P of posting lists. The term "football" is contained in the Web pages whose docID is 50, 100, 500, and whose URL address is reported in table U

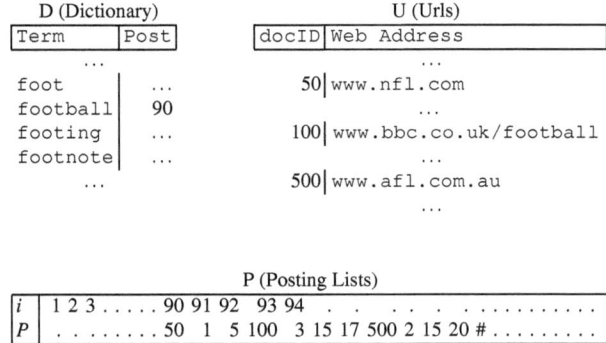

succinctly identified by integers, called *docID*s, which have been assigned during the crawling process. The dictionary of terms is also stored in a table which contains some satellite information and the pointers to the posting lists. Of course, the storage of all terms in the documents impacts the total space required by the index. Previously, software engineers preferred to restrict the indexed terms to only the most significant ones; nowadays, search engines index essentially all terms extracted from the parsed documents because advances in data compression allowed engineers to squeeze terms and docIDs in reduced space and still guarantee fast query responses. Actually, search engines also store the *positions* of all term occurrences in each indexed document because this information is used to support phrase searches and to estimate the relevance of a document with respect to a query, based on the distance between the query terms in that document. It is evident that such a refined set of information has huge size and thus necessitates sophisticated compression techniques. The literature reports several studies about this issue, nonetheless the actual compressors adopted by commercial search engines are mostly unknown.

Figure 5.8 illustrates the structure of an inverted list. Each term t ("football" in the example) has associated a subarray of P which stores, in order, the docID of a document d containing term t (the first document in the example is 50), the number of times that t occurs in d (1 in the example), the position in d of each of these term occurrences (position 5 in the example). The posting list of t ends with a terminator #. From the figure we notice that the term $t =$ football is contained in document 50 at one single position, i.e., 5; in document 100 the term occurs in three positions (15, 17 and 25); in document 500 it occurs in two positions (15 and 20). It is convenient to store the docIDs of each posting list in increasing order (50, 100, 500 in the example) because this reduces the space occupancy and the time required to solve future queries. In this case each docID can be stored as the *difference* with respect to its preceding docID. The same method can be used to succinctly store the positions of the occurrences of term t in document d. So, in the posting list of Fig. 5.8, we can represent the sequence of docIDs 50 100 500 as 50 50 400: the first 50 is exactly represented, since it has no preceding docID, whereas we have $100 - 50 = 50$ and $500 - 100 = 400$ for the next two docIDs. By also inserting the occurrences of the term, encoded similarly, the compressed posting-list of the term "football" becomes:

$$50\ 1\ 5\ 50\ 3\ 15\ 2\ 8\ 400\ 2\ 15\ 5\ \#.$$

The original posting list is easily reobtained from the compressed one by a simple sequence of additions. We are speaking about "compression" because the obtained numbers are smaller than the original ones, so we can squeeze the total space usage by means of appropriate integer coders that produce short bit sequences for small integers.

As we anticipated above, the order of the docIDs in the posting lists is important also to efficiently answering queries that consist of more than one keyword. Imagine that a user has formulated a query with two keywords, t_1 and t_2 (the extension to more keywords is immediate). Solving this query consists of retrieving the posting lists L_1 and L_2 of docIDs referring to t_1 and t_2, respectively. As an example, take $L_1 = $ 10 15 25 35 50 ...# and $L_2 = $ 15 16 31 35 70...#, where we are assuming to have already decompressed the lists. Now the problem is to identify the documents that contain both t_1 and t_2 (namely, the elements in common to both the two posting lists). The algorithm is deceptively simple and elegant; it consists of scanning L_1 and L_2 from left to right comparing at each step a pair of docIDs from the two lists. Say $L_1[i]$ and $L_2[j]$ are the two docIDs currently compared, initially $i = j = 1$. If $L_1[i] < L_2[j]$ the iterator i is incremented, if $L_1[i] > L_2[j]$ the iterator j is incremented, otherwise $L_1[i] = L_2[j]$ and thus a common docID is found and both iterators are incremented. If we let n_1 and n_2 be the number of elements in the two lists, we can realize that this algorithm requires time proportional to $n_1 + n_2$. A each step, indeed, we execute one comparison and advance at least one iterator. This cost is significantly smaller than the one required to compare each element of L_1 against all elements of L_2 (which is $n_1 \times n_2$), as would happen if the lists were not sorted. Since the values of n_1 and n_2 are on the order of some hundreds of thousands (or even more for the common terms), the latter algorithm would be too slow to be adopted in the context of a search engine answering millions of queries per day.

5.4.4 Evaluating the Relevance of a Page

Since user queries consist of a few keywords, the number of pages containing these keywords is usually huge. It is thus vital that a search engine sorts these pages and reports the most "relevant" ones in the top positions to ease their browsing by the user. However, an accurate characterization of what is the "relevance" of a Web page has a high degree of arbitrariness. Nonetheless, this is probably the most important step in modern search engines, so that a bunch of sophisticated algorithms have been proposed to efficiently quantify that relevance. This step goes under the name of *ranking*, and its solution represents the main point of distinction between the major known search engines. It is indeed not exaggerated to affirm that one of the key ingredients that enabled Google to achieve its enormous popularity was its algorithm for ranking the pages shown to the user, called *PageRank* (see below for details).[7]

Nowadays the relevance of a page p is measured by combining several parameters: the type and the distribution of the occurrences of the query-terms in p, the *position* of p in the Web graph and its interconnections with other pages, the frequency with which Web users visit p as a result of a user query, and many other factors, not all revealed by search engines. In particular, it is known that Google uses about 100 parameters! We present below the two most important measures of relevance for Web pages known in the literature, prefacing them with some considerations that will allow us to understand their inspiring motivations.

It is natural to think that the relevance of a term t for a document d depends on the frequency $TF[t, d]$ of occurrence of t in d (called *Term Frequency*), thus, on the *weight* that the author of d has assigned to t by repeating this term several times in the document. However, considering only the frequency may be misleading because, for example, the articles and prepositions of a language are frequent in texts without characterizing them in any way. Thus, it is necessary to introduce a correction factor which also takes into account the *discriminative power* of a term which is very low for secondary linguistic elements. However, the situation is more complicated, since a term like "insect" may be discriminant or not, depending on the collection of documents in its entirety: "insect" is unusual and probably relevant for a collection of computer-science texts, whereas it is obvious and therefore irrelevant for a collection of entomology texts. It is therefore crucial to consider the rarity of a term in the collection by measuring the ratio between the number ND of documents in the collection and the number $N[t]$ of documents containing the term t. The rarer the term t, the larger the ratio $ND/N[t]$, and thus t is potentially more discriminative for the documents in which it is contained. Usually this ratio is not directly used to estimate the discrimination level of t, but it is mitigated by applying the logarithmic function. This defines the parameter $IDF[t] = \log_2(ND/N[t])$, which is called the *Inverse Document Frequency*. In this way, the measure is being not too sensitive to small variations in the value of $N[t]$. On the other hand, it is not always true that a rare word is very relevant for the document d. For example, the presence of the word may be caused by a typing error. Therefore, term frequency and inverse document frequency are combined to form the so-called TF-IDF measure of relevance of a term in a document. This combination was proposed in the late 1960s and is given by the formula $W[t, d] = TF[t, d] \times IDF[t]$. Notice that if t is, say, an article, it appears frequently in almost all the documents in the collection. Thus, its ratio $ND/N[t]$ is very close to the value 1 and its logarithm is close to the value 0, thus forcing a small value of $W[t, d]$. Similarly, a term typed incorrectly will have a

---

[7]Google trusts so much in its ranking algorithm that it still shows in its homepage the button "I'm feeling lucky" that immediately sends the user to the first ranked page among the results of her query.

small value for *TF*, thus forcing again a small value of $W[t,d]$. In both cases then the term relevance will be correctly evaluated as not significant. Numerous linguistic studies have corroborated the empirical validity of the TF-IDF weight which is now at the basis of any information retrieval system.

The first generation of search engines, such as Altavista, adopted the TF-IDF weight as a primary parameter to establish the importance of a Web page and sorted the query results accordingly. This approach was effective at the time in which the Web content was mainly restricted to government agencies and universities with authoritative pages. In the mid-1990s the Web was opened to the entire world and started to become a huge "shopping bazaar", with pages composed without any control in their content. All this led some companies to build "rigged" pages, namely pages that contained, in addition to their commercial offerings, also the set of properly concealed keywords that frequently appeared in user's queries. The net aim was to promote the relevance of these pages, even in other contexts. Thus, it was evident that the textual TF-IDF weight alone could not be used to assess the importance of a page, but it was necessary to take into account other factors specific to the Web graph.

Since the mid-1990s several proposals came from both academia and industry with the goal of exploiting the links between pages as a *vote* expressed by the author of a page p to the pages linked by p. Among these proposals, two ranking techniques gave rise to the so-called *second generation search engines*: the first technique, called *PageRank*, was introduced by the founders of Google, Larry Page and Sergey Brin, and the second technique, called *HITS* (Hyperlink Induced Topic Search), was introduced by Jon Kleinberg when he was at IBM. In PageRank each page is assigned with a relevance score which is independent of its textual content, and thus of the user query, but depends on the Web-graph topology. In HITS each page is assigned with two relevance scores which depend on the topology of a subgraph selected according to the user query. Although very different, PageRank and HITS have two common features: they are defined recursively, so the relevance of a page is computed from the relevance of the pages that are linked to it; they involve computations on very large matrices derived from the structure of the Web graph, so they need sophisticated mathematical tools (readers less familiar with linear algebra can jump to Sect. 5.4.6).

5.4.5 Two Ranking Algorithms: PageRank and HITS

PageRank measures the relevance of a page p according to its "popularity" in the Web graph, which in turn is computed as a function of the number and origin of links that point to p. In mathematical terms, the popularity $R(p)$ is computed as the probability that a user will reach page p by randomly walking over the Web graph, traversing at each step one of the links outgoing from the currently visited page, each selected with equal probability. Let $p_1, \ldots, p_k$ be the pages having at least one link to p, and let $N(p_i)$ be the number of pages linked by p_i (i.e., the number of outgoing links from p_i in the Web graph). The basic formula for the calculation of $R(p)$ is the following:

$$R(p) = \sum_{i=1...k} (R(p_i)/N(p_i)). \qquad (5.2)$$

Notice that only the pages having a link to p directly contribute to the value of $R(p)$, and moreover, this contribution is proportional to the relevance of these pages scaled by their number of outgoing links. The ratio underlying this formula is that if a page p_i with a certain relevance $R(p_i)$ points to p, it increases the popularity of p, but this increment should be equally shared among all the pages to which p_i points.

It is evident that the formula is recursive, and its computation presents some technical problems because it requires us to specify the initial value of $R(p)$, for all pages p, and to indicate how to deal with pages that do not have either incoming or outgoing edges. To address these two issues, we consider a slightly different formula that introduces a correction factor taking into account the possibility that a user leaves the link-traversal and jumps to a randomly chosen page in the Web graph. This change allows the random walker to not remain stacked into a page that has no outgoing links, or to reach a page even if it has no incoming links. Therefore, the formula becomes:

$$R(p) = d \sum_{i=1...k} (R(p_i)/N(p_i)) + (1-d)/n, \qquad (5.3)$$

where n is the number of pages collected by the crawler and indexed by the search engine, and d is the probability of continuing in the link-traversals, whereas $(1-d)$ is the complement probability of jumping to a randomly chosen page in the crawled graph. In the extreme case that $d = 0$, all pages would obtain the same relevance $R(p) = 1/n$, while in the case of $d = 1$ the relevance $R(p)$ would entirely depend on the structure of the Web graph and it would show the problems mentioned above. Experiments have suggested taking $d = 0.85$, which actually attributes more importance to the relevance that emerges from the structure of the Web.

The real computation of $R(p)$, over all pages p, is performed by resorting to matrix operations. We have already seen that the Web graph may be represented by a matrix W whose powers W^k indicate the number of paths of length k between pairs of nodes in the graph (see box *Adjacency matrix and paths of a graph*). For this reason, we introduce a matrix Z of size $n \times n$, whose elements have value $Z[i, j] = d \times W[i, j] + (1-d)/n$. The value $Z[i, j]$ represents the probability that a random walker traverses the link from p_i to p_j, while the matrix-powers Z^k represent the probability that paths of length k are traversed by that random walker.

We can also represent the relevance of the pages in vector form: $R[i]$ is the relevance of page p_i; and use the notation $R_k[i]$ to denote the relevance of page p_i

after k iterations of the algorithm. At this point we can compute the configurations for all R_k as:

$$R_1 = R_0 \times S, R_2 = R_1 \times Z = R_0 \times Z^2, \ldots, R_k = R_0 \times Z^k. \qquad (5.4)$$

This formula is related to a deep mathematical theory known as *Markov's chains*, which is, however, too difficult to be discussed here. We note that this theory guarantees that the limit value for $R_k[i]$, when $k \to \infty$, equals the probability that a random walker traverses page p_i, and it also guarantees that this limit value does not depend on the initial conditions R_0, which can then be assigned arbitrarily.

Obviously, the calculation indicated in Eq. (5.4) is not trivial due to the size of the matrices involved, which are indeed huge since they consist nowadays of hundreds of billions of pages and hence have a total size of at least 25×10^{20} elements! However, the computation of R_k is simplified by the fact that we do not need to care about the exact values of its elements, since it suffices to determine only their order: if $R(p_1) > R(p_2)$, then page p_1 is more important than page p_2. Therefore, we can stop the above computation whenever the values of R_k's components are sufficiently stable and their order can be determined with some certainty. Experimental tests showed that about 100 iterations suffice.

We conclude the discussion on PageRank by recalling that it induces an ordering among pages which is a function only of the graph structure and thus it is independent of the user query and page content. Therefore, PageRank can be calculated at the indexing phase, and deployed at query time in order to sort the result pages returned for a user query. Many details on the current version of PageRank are unknown, but numerous anecdotes suggest that Google combines this method with TF-IDF and a 100 other minor parameters extracted automatically or manually from the Web.[8]

Let us study now the HITS algorithm, which is potentially more interesting than PageRank because it is *query dependent*. For a given query q, it retrieves first the set P of Web pages that contain all query terms, and then it adds those pages that point to or are pointed to by pages in P. The resulting collection is called the *base set* and contains pages that are related to q either directly (because they contain the query terms) or indirectly (because they are connected to a page in P). The situation is shown in Fig. 5.9a.

A (sub-)graph is then built by setting the pages in the base set as nodes and the links between these pages as edges of the (sub-)graph. For each node, we calculate two measures of relevance, called *authority* and *hubness* scores. The first score, denoted with $A(p)$, measures the authoritativeness of page p relative to the query q.

---

[8]In a recent interview, Udi Manber (VP Engineering at Google) revealed that some of these parameters depend on the language (ability to handle synonyms, diacritics, typos, etc.), time (some pages are interesting for a query only if they are fresh), and templates (extracted from the "history" of the queries raised in the past by the same user or by her navigation of the Web).

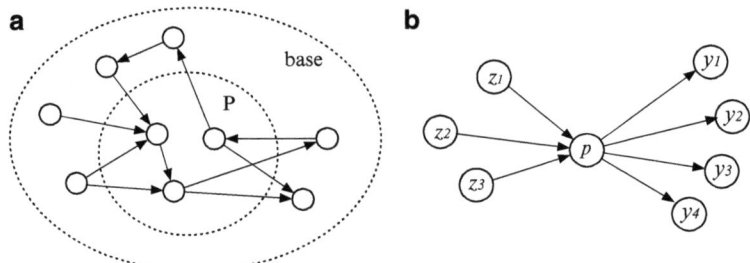

Fig. 5.9 (a) The set P of pages that contain the terms of a query q, and its base set; (b) the pages z_i and y_i that contribute to determine respectively the *authority* and the *hubness* score of a page p

The second score, denoted with $H(p)$, measures how much the p's content is a good survey for the query q (i.e., a directory that points to many authoritative pages on the subject). This way, a page p having a large value of $A(p)$ is an *authority* for q, while a large value of $H(p)$ implies that p is a *hub* for q. Computing these two measures follows their intuitive meanings: a page p is a good hub (and, thus, the value $H(p)$ is large) as p points to many authoritative pages; a page p is a good authority (and, hence, the value of $A(p)$ is large) as more good hubs point to p. We can formalize the previous insights with the following two formulas:

$$A(p) = \sum_{i=1...k} H(z_i);$$
$$H(p) = \sum_{i=1...k} A(y_i), \qquad (5.5)$$

where $z_1, \ldots, z_k$ denote the pages that point to p, and $y_1, \ldots, y_h$ denote the pages pointed to by p (see Fig. 5.9b). Similarly to what was done for the PageRank, we can compute the two scores by resorting matrix computations. We then define the adjacency matrix B of the graph induced by the base set, and we compute the vectors A and H with the formulas (5.5) via matrix computations involving B. These computations are similar to the ones performed for PageRank with two essential differences. The first concerns the size of the matrices, which is now moderate since B usually consists of only a few thousands of nodes. The second is that the calculation has to be executed on-the-fly at query time because B is not known in advance, and thus the values of A and H cannot be precalculated. This represents a strong limitation for the application of this method on the Web, and in fact HITS was originally proposed for search engines operating on small collections of documents and for a limited number of users (e.g., on a company intranet). Another limitation of this approach resides in its small robustness to *spam* (see Sect. 5.4.6); for this reason in the literature this issue got some attention with many interesting proposals.

5.4.6 On Other Search Engine Functionalities

Among the other operations that a search engine is called to perform, the presentation to a user of the results of her query has great importance. Search engines show, for each result page, a short textual fragment, known as a *snippet*, which represents the context surrounding the query terms in that page. Other search engines offer also the possibility of viewing a copy of a result page as it was retrieved by the crawler. This is particularly useful whenever the link returned for a result page is broken, due to the fact that this page was removed from the Web since it was crawled. The copy of this page can thus be useful to retrieve the indexed page, even if it is no longer present in the Web.

We emphasize that the results of search engines are sometimes corrupted with sophisticated techniques, which fraudulently increase the relevance of certain pages to let them appear among the first results for a user query. This is known as *spamming* and consists of constructing proper subgraphs of the Web that artificially increase the relevance of the fraudulent pages. Other forms of spamming are more subtle. One technique, known as *cloaking*, is adopted by fraudulent servers to mask the real content of their pages and thus make them the result of queries which are not related to their content. The cloaking idea makes servers return to the search engine some good content taken, for example, from Wikipedia at each crawling request. If the artifact is relevant for a user query, the search engine will then display a snippet appropriate and interesting for the user and referring to a Wikipedia page. However, the page that appears to the user after clicking on the link shown in the snippet is totally different and possibly contains irrelevant (if not offensive) content.

Certainly, an exhaustive discussion on spamming methods cannot be addressed in this short chapter. However, it is interesting to note that spamming has a large reach, as it is estimated that more than 20 % of Web pages consist of artifacts that endanger the reputation and usefulness of search engines. In all their phases, consequently, search engines adopt sophisticated anti-spam algorithms to avoid the gathering, the indexing and the reporting of these artifacts. As for all the previous problems, the solutions currently employed by commercial search engines are only partially revealed, in part to make the job of spammers more difficult.

We finally remark that the goal of search engines is moving toward the identification of user intentions hidden behind the purely syntactic composition of their query. This explains the proliferation of different methods that cluster the query results on the screen (started by the search engine Vivisimo.com), that integrate different sources of information (news, Wikipedia, images, video, blogs, shopping products, and so on), and that possibly provide suggestions for the composition of refined queries (Google Suggest and Yahoo! Search Suggest are the most notable examples). In addition, search engines have to tackle the fact that users are moving from being *active agents* in the search process to becoming more and more *passive spectators*: advertising, suggestions, weather forecasts, friends on-line, news, and so on, are all information that we probably set as interesting in some personal record or alerts, or that the search engines have in some way inferred as interesting for us given our Web life. All of this information already appears, or will appear more and more frequently in the near future, on our screens as a result of a query, or on our email readers, our personal pages on iGoogle or MyYahoo!, or even on applications in our smartphones.

Many other features of search engines deserve to be carefully studied and discussed. An interested reader can find many inspiring references in the review of the literature at the end of this chapter.

5.5 Towards Semantic Searches

Although Web search engines are fairly new, researchers and software engineers achieved during the last two decades significant improvements in their performance. These achievements identified many other interesting avenues of further research which should lead in the near future to implementing more efficient and effective Information Retrieval (IR) tools. In fact, although the algorithms underlying the modern search engines are much sophisticated, their use is pretty much restricted to retrieving documents by keywords. But, clearly, users aim for much more!

Keyword-based searches force users to abstract their needs via a (usually short) sequence of terms; this process is difficult and thus error prone for most of Web users, who are unskilled. It would surely be more powerful to let users specify their needs via natural-language queries: such as "Will it rain in Rome within the next three hours?", and get more precise answers than just an ordered list of pages about Rome, or a Web page about the weather in Rome, such as "yes, it will rain on the coast". Interestingly enough, this is not just a matter of ranking; we are actually asking the search engine to understand the *semantics* underlying the user query and the content of the indexed pages. Some interesting research is actually going on, trying to address these issues by adding metadata to pages in order to better describe their content (known as the *Semantic Web*, and as the *Resource Description Framework*), or by adding some structure to pages in order to simplify the automatic extraction of useful information from them (known as *Open Data*), or by developing powerful Natural Language Processing techniques that better interpret short phrases up to long documents. This last research line has led to some interesting results that we will sketch briefly in this final part of the chapter.

The typical IR approach to indexing, clustering, classification, mining and retrieval of Web pages is the one based on the so-called *bag-of-words paradigm*. It eventually transforms a document into an array of terms, possibly weighted with TF-IDF scores (see above), and then represents that array via a highly dimensional point in a Euclidean space. This representation is purely syntactical and unstructured, in the sense that different terms lead to different and independent dimensions. Co-occurrence detection and other processing steps have been thus proposed to identify the existence of synonymy relations, but everyone is aware of the limitations of this approach especially in the expanding context of short (and thus poorly composed) documents, such as the snippets of search-engine results, the tweets of a Twitter channel, the items of a news feed, the posts on a blog, etc.

A good deal of recent work attempts to go beyond this paradigm by enriching the input document with additional *structured annotations* whose goal is to provide a contextualization of the document in order to improve its subsequent "automatic interpretation" by means of algorithms. This general idea has been declined in the literature by identifying in the document short and meaningful sequences of terms, also known as *entities*, which are then connected to unambiguous topics drawn from a catalog. The catalog can be formed by either a small set of specifically recognized types, most often People and Locations (also known as *named entities*), or it can consists of millions of generic entities drawn from a large knowledge base, such as Wikipedia. This latter catalog is ever-expanding and currently offers the best trade-off between a catalog with a rigorous structure but with low coverage (like WordNet or CYC), and a larger catalog with wide coverage but unstructured and noisy content (like the whole Web).

To understand how this annotation works, let us consider the following short news: "Diego Maradona won against Mexico". The goal of the annotation is to detect "Diego Maradona" and "Mexico" as significant entities, and then to hyperlink them with the Wikipedia pages which deal with the two topics: the former Argentinean coach and the Mexican football team. The annotator uses as entities the anchor texts which occur in Wikipedia pages, and as topics for an entity the (possibly many) pages pointed in Wikipedia by it, e.g., "Mexico" points to 154 different pages in Wikipedia. The annotator then selects from the potentially many available mappings (entity-to-topic) the most pertinent one by finding a collective agreement among all entities in the input text via proper scoring functions. There exist nowadays several such tools[9] that implement these algorithmic ideas and have been successfully used to enhance the performance of classic IR-tools in classification and clustering applications. Current annotators use about eight million entities and three million topics.

We believe that this novel entity-annotation technology has implications which go far beyond the enrichment of a document with explanatory links. Its most interesting benefit is the structured knowledge attached to textual fragments that leverages not only a *bag of topics* but also the powerful *semantic network* defined by the Wikipedia links among them. This automatic *tagging* of texts mimics and automates what Web users have done with the advent of Web 2.0 over various kinds of digital objects such as pages, images, music and videos, thus creating a new parallel language, named "*folksonomy*". This practice has made famous several sites, such as Flickr, Technorati, Del.icio.us, Panoramio, CiteULike, Last.fm, etc. Topic annotators could bring this tagging process to the scale of the Web, thus improving the classification, clustering, mining and search of Web pages, which then could be driven by topics rather than keywords. The advantage would be

---

[9]See, for example, TAGME (available at tagme.di.unipi.it), and WIKIPEDIA MINER (available at http://wikipedia-miner.cms.waikato.ac.nz/).

the efficient and effective resolution of ambiguity and polysemy issues which often occur when operating with the purely syntactic bag-of-words paradigm.

Another approach to enhance the mining of Web pages and queries consists of extracting information from *query-logs*, namely the massive source of queries executed by Web users and their selected results (hence, pages). Let us assume that two queries q_1 and q_2 have been issued by some users and that they have then clicked on the same result page p. This probably means that p's content has to do with those two queries, so that they can be deployed to extend p's content, much as we did with anchor texts of links pointing to p in the previous sections. Similarly, we deduce that queries q_1 and q_2 are probably correlated and thus one of them could be *suggested* to a user as a refinement of the other query. As an example, Google returns for the two queries "iTunes" and "iPod" the page http://www.apple.com/itunes/ as the first result. So we expect that many users will click on that link, thus inferring a semantic relation between these two queries.

There are also cases in which the same query q might lead users to click on many different page results; this might be an indication that either those pages are similar or that q is polysemous. This second case is particularly important to be detected by search engines because they can then choose to adopt different visualization forms for the query results in order to highlight the various *facets* of the issued query and/or *diversify* the top answers with samples of pertinent topics. As an example, let us consider the query "eclipse", which could be issued by a user interested in astronomical events, or in the software development framework, or in a plane model (Eclipse 500), or in a Mitsubishi car. So the query-log will contain many pages which are semantically connected to the query "eclipse", all of them pertinent with its various meanings.

It is therefore clear at this point that the analysis of all queries issued to a search engine and of the clicks performed by their issuing users can lead us to construct a huge graph, called the *query-log graph*, which contains an impressive amount of semantic information about both queries and pages. The mining of the structure and content of this graph allows us to extract impressive amounts of useful knowledge about the "folksonomy" of Web searches, about the community of Web users and their interests, about the relevant pages frequently visited by those users and their semantic relations. Of course, a few clicks and searches are error prone, but the massive amounts of issued queries and user clicks made every day by the Web community make the information extractable from this graph pretty robust and scalable to an impressive number of topics and Web pages. It is evident that we are not yet at a full understanding of the content of a page, neither are we always able to disambiguate a query or fully understand the user intent behind it, but we are fast approaching those issues!

5.6 Bibliographic Notes

Web search is a complex topic which was worth thousands of scientific papers in the last three decades. In this section we report the books and some scientific articles that have advanced the history of this fascinating and rapidly evolving field of research. These references offer a good and accessible treatment of the various topics dealt with by this chapter.

The book written by Witten et al. [113] contains a general study of the characteristics of search engines and implications on their use. This very clear text represents a great source of knowledge that does not enter much into algorithmic details. Two recent books by Manning et al. [75], and by Baeza-Yates and Ribeiro-Neto [7], describe the basics of Information Retrieval, whereas the book by Chakrabarti [17] provides a complete introduction on the gathering and analysis of collections of Web pages. Finally, the book by Witten et al. [112] is a fundamental reference for what concerns the organization and the processing of massive data collections (not necessarily formed by Web pages).

As far as papers are concerned, we mention the publications by Hawking [57, 58] and Lopez-Ortiz [74], which offer two ample and clearly written surveys on the algorithmic structure of search engines. On the other hand, the paper by Zobel and Moffat [116] concentrates on indexing data structures, describing in much detail Inverted Lists. Two historical papers are the ones published by Broder et al. [12] on the analysis of the Web graph, and by Brin and Page [11] on PageRank, thus laying down the seeds for Google's epoch.

As far as Web tools are concerned, we point out the papers by Fetterly [42], who deals with spamming and adversarial techniques to cheat crawlers, by Baeza-Yates et al. [8], who describe the scenario of semantic search engines, and by Silvestri [101], who surveys the use of query-logs in many IR applications. Finally, we mention the papers by Ferragina and Scaiella [39] and Scaiella et al. [98], which review the literature about "topic annotators" and their use in search, classification and clustering of documents. We conclude this chapter by pointing out to the readers Google's blog by Alpert et al. [5], in which these engineers claimed that Google crawled and indexed in July 2008 about one trillion pages.

Acknowledgements We would like to thank Fabrizio Luccio, who contributed to the writing of the Italian version of this chapter for Mondadori.

Chapter 6
Algorithms for Secure Communication

Alberto Marchetti-Spaccamela

Abstract The design of algorithms for sending confidential messages (i.e. messages that no one can read, except the intended receiver) goes back to the beginning of our civilization. However, before the widespread of modern computers, cryptography was practiced by few people: soldiers, or diplomats, or scientists fascinated by the problem of confidential communication. Cryptography algorithms designed in the past were ingenious transformations but were lacking a sound mathematical basis. Recently, the development of computers and of the Internet has opened up new applications of cryptography in business and society. To answer these needs, new algorithms have been developed that use sound mathematical techniques and have produced surprising results, which have opened up impressive possibilities that were considered unrealistic before. We will see examples of algorithms that use modular arithmetic (in which operations are performed modulo an integer) that are based on using functions that are easy to compute but difficult to invert.

6.1 Introduction

Cryptography has a long and fascinating history that started when the military and diplomats of the first organized states recognized the importance of message confidentiality in the event that the bearer of the message was captured by the enemy. To achieve this, the original M, the plaintext message brought by a messenger to a recipient, must be transformed into a new text Y, the message encrypted or encoded. When the recipient of the message receives Y he should be able to reconstruct the original text, but at the same time the encoded message Y must be incomprehensible

---

A. Marchetti-Spaccamela (✉)
Dipartimento di Ingegneria Informatica, Automatica e Gestionale, Sapienza Università di Roma, via Ariosto 25, 00185 Roma, Italy
e-mail: alberto.marchetti-spaccamela@uniroma1.it

to anyone else who came into possession of the message (for example, the enemy that took it from the captured messenger).

The advantages of secure methods for communications are obvious and therefore it is not surprising that we found ways to obscure communications at the beginning of our civilization that can be considered the first examples of cryptography. For example, we found ways of obscuring texts in the Egyptian hieroglyphics and in Assyrian cuneiform scripts used by the Babylonians. Methods to hide the meaning of texts are also mentioned in the Indian book *Kama Sutra* among the arts that men and women should learn. In this case, knowledge is not for war but allows lovers to communicate without being discovered. In particular, the 44th and 45th of the 64 arts of the book are:

> 44. The art of understanding writing encrypted and spelling in particular.
>
> 45. The art of speaking by changing the form of words. For this art there are several possibilities: some change the start or end of words, others add unnecessary letters between every syllable of a word.

We will see in the following that, until the last century, cryptography can be considered an art practiced by few people: professionals whose only practical purpose was in the military and diplomatic, or scientists fascinated by it. Over the last 50 years, there has been a rapid growth of cryptography; the pervasive development of computers and of the Internet has opened up new applications in business and society. In fact, on one hand the development of powerful computers has made it possible to implement sophisticated algorithms that require complex calculations to encode and decode; on the other hand, the development of the Internet has opened up new prospects for use, such as electronic commerce and business. As a consequence, a new approach to cryptography has been developed that uses sophisticated mathematical techniques and methodologies.

The practical consequence is that nowadays cryptographic algorithms are used by the layman with no specific knowledge and in most cases without knowing it.[1] For example, when we buy a book or a service over the Internet using our credit card credentials, we generally use cryptographic protocols that make the communicated data unintelligible to a possible intruder that is able to get the exchanged data.

We finally notice that the commercial development of the Internet has given rise to new problems and new fields of application in addition to the traditional secure encryption of messages. In the final paragraph of this chapter we shall refer to one of these, the digital signature, showing how you can sign documents electronically to avoid falsification of the signature itself and ensure their authenticity.

---

[1]In particular, when buying a product on the World Wide Web and providing the payment information, the web protocol automatically changes from http to https. The https protocol is able to realize cryptographic encodings that are not possible with the http protocol, the one usually used for browsing the Web; the final letter of https means secure.

6.2 A Brief History of Cryptography

The history of cryptography is fascinating and long. In the following we will not go into details and we limit our discussion to some useful elements for understanding the rest of this chapter; we refer to [64, 103] for a detailed presentation of the history of cryptography.

Most cryptographic codes that were proposed in the past based on the exchange or replacement of characters of the text. An example is when we exchange the mutual position of the first pair of characters, of the second pair and so on, thus writing "esrcte" instead of "secret". Another example is specular writing, as used by Leonardo da Vinci in his notebooks.[2]

6.2.1 Monoalphabetic Substitution Codes

Most codes are based on the replacement of characters of the original text with other characters. The most famous of these methods proposed in the past is the Caesar cipher, in which a simple transformation of the text by sliding the alphabet is performed. In particular, Suetonius (*Lives of the Caesars*, 56) reports that

> if Caesar had something secret to communicate, he wrote in cipher, changing the order of the letters of the alphabet, so that we could not understand any word. If anyone wanted to decipher the text and obtain its meaning he must substitute the fourth letter of the alphabet, namely D, with A, and so on with the others.

Caesar's code can be generalized using other substitution orderings (i.e., scrolling through the alphabet by a different number of positions instead of three positions as reported by Suetonius). Note that, given a message M, if we encode it by scrolling the alphabet by five positions instead of three we get different encrypted messages. In other words, the encoding obtained by applying Caesar's code depends both on the message M to be encoded and on a key which indicates the number of positions in the sliding of the alphabet. In particular, in the previous citation of Suetonius, 3 is the key used.

One major limitation of Caesar's code is that once we know the method, it is relatively easy to decode messages, even if we do not know the key. In fact, since there are only 25 possible ways of scrolling the alphabet, it is possible to try them all to finally know the original message. Motivated by this, a more sophisticated approach is to use a completely arbitrary substitution of alphabetic character, such as that described in the following table that assigns the letter "M" to the letter "A", and so on.

---

[2]Specular writing is handwriting that is written from right to left (left-handed writing), and therefore can be only deciphered by means of a mirror.

Table 6.1 A substitution code

A	B	C	D	E	F	G	H	I	J	K	L	M	N	O	P	Q	R	S	T	U	V	W	X	Y	Z
M	F	H	A	I	V	R	N	O	B	L	K	E	G	P	X	W	J	S	D	U	T	Q	Y	C	Z

Table 6.2 A substitution code that is easy to break

A	B	C	D	E	F	G	H	I	J	K	L	M	N	O	P	Q	R	S	T	U	V	W	X	Y	Z
A	B	C	D	E	F	G	H	I	J	K	L	M	N	O	P	R	Q	S	T	U	V	W	X	Y	Z

So if we want to encode the first verse of Shakespeare's Macbeth

When shall we three meet again?

with the code described in Table 6.1, we get the following encoded message:

QNIG SNMKK QI DNJII EIID MRMOG?

In this case, the number of different encodings of the same text corresponds to all possible permutations of the characters of the alphabet. However, it is easy to realize that not all possible permutations make it difficult to analyze the message. For example, the messages encoded by using the correspondence described in Table 6.2 (in which only the letters R and Q are permuted) are easily understood.

Caesar's cipher and codes like the one given in Table 6.1 are monoalphabetic codes (since they replace one character with another character). They were a breakthrough in their time because they showed the possibility of making obscure messages relatively easily, thus obtaining a strategic advantage in military communications. These considerations motivated the proposal of many other algorithms to keep information secret, leading to the birth of a new art, the art of cryptography.

At the same time, the proposed new encryption methods motivated the study of the weaknesses of the methods proposed with the opposite goal of breaking the code. We can imagine this process as a struggle between the inventors of secret codes and the codebreakers (the opponents), who aim at deciphering the message (or "breaking" the code), managing to decode encrypted messages in a relatively short time.

The encrypted text using the replacement products and other techniques proposed in the past can be easily attacked using statistical information on the frequency of characters in the original text. In particular, in the ninth century, Arab mathematician Al-Kindi's contribution was to show that in a text alphabetic characters are present with a frequency different from character to character, and that such frequencies are significantly different depending on the language. Moreover, frequencies do not depend significantly on the text itself; namely, given a sufficiently long text, character frequencies are relatively stable and depend only on the language.[3]

---

[3]The text must be long enough to meet the average frequencies of characters. In fact, it is not difficult to write short sentences in which the frequency is very different.

For example, the letters A, T appear in English with an average frequency of about 8–9 %; the letter E with a frequency above 12 %; the letters J, K, Q, V, X and Z have a frequency less than 1 %. This implies that in a sufficiently long ciphertext, if we use the substitution character described in Table 6.1, the most frequent characters in the ciphertext will be the characters M, I and D. Al-Kindi showed how knowledge about the frequency of characters allowed one in a relatively easy way to reconstruct the original message from a long enough ciphertext. This observation has led to monoalphabetic codes not being considered safe with respect to a crafty attacker.

Leon Battista Alberti, the famous Renaissance architect, plays a key role in the history of cryptography for his proposal of a new type of cryptographic algorithm, known as polyalphabetic codes. Alberti's idea was to use different substitutions for different parts of the message. For example, we can use three different monoalphabetic substitution codes and apply the first one to the first character of the message, the second one to the second character, and so on (starting again with the first code at the fourth character).

It is clear that if we use only three monoalphabetic codes we cannot expect to obtain an encoded message that is resilient to the analysis based on the frequencies of characters. In fact, knowing that the same monoalphabetic code is used for the first, fourth, seventh, etc. character, then frequency analysis can be applied to this part of the plaintext. However, Alberti's method allows one to use many different codes. If we increase the number of monoalphabetic codes, the obtained polyalphabetic code becomes more reliable but at the cost of increasing the complexity of encoding and decoding because we need many monoalphabetic substitution tables. To overcome this drawback Alberti devised a tool to facilitate these operations consisting of several disks that could rotate independently.

6.2.2 Polyalphabetic Substitution Codes

Alberti's idea was refined and improved in several ways. From a historical point of view, the best-known and most widely used method is named after Blaise de Vigenère, a French diplomat who presented his method to Henry II of France in 1586. However, the method proposed by Vigenère is a simplification algorithm that G. Bellaso published in 1553 in his book *La Cifra* (*The Cipher*). Vigenère almost certainly did not know the work of Bellaso and so the two cryptographers had the same idea almost simultaneously; however, even if the primogeniture of the idea is undoubtedly by Bellaso, for historical reasons, the method is known as the method of Vigenère.

The main innovation of Bellaso and Vigenère was to simplify the implementation of Alberti's proposal by using a secret word that controls which of several monoalphabetic substitution codes must be adopted. For example, suppose that "robin" is the secret word, then we note that "r", the first character of the secret word, is the 18th letter of the alphabet; therefore the Vigenère method encodes the first character of text using a Caesar cipher with displacement of 17 positions. Since the second character of the secret is the letter "o" (the 15th character of the alphabet), the second character of the text is encrypted using a Caesar cipher with a shift of 14 positions. The method proceeds in a similar way: the letter "b" is the third character of the secret, so the third character of the text is encoded with a Caesar cipher in which we carry out the displacement of one position ("b" is the second character of the alphabet). In this way we can encode the first five characters of the text (as many as there are characters in the secret word). To encode the rest of the message we reiterate the method, so we encode the sixth character of the text (and, later, the 11th, the 16th, etc.) using a Caesar cipher with a shift of 17 positions, as done with the first letter.

Note that the transformation of the original text in the text being encrypted is done using an algorithm and a key ("robin" in the previous example) that specifies how the original text is transformed. Clearly, using the method on the same text using different keys allows one to obtain encrypted messages that are completely different. In particular, the encrypted message is incomprehensible to an opponent unfamiliar with the key, even if he knows that the message is encrypted using Vigenère's algorithm. In this way, the same algorithm can be used to communicate with more people using a different key for each communication. It should be also clear that if we use as a secret a long text instead of a single word, then the frequency analysis that allows us to break Caesar's code becomes much more difficult (in fact, for a long time the Vigenère code was named the unbreakable code).

The simultaneous use of different monoalphabetic codes implies that the frequency of the characters in the ciphertext depends on the original text and several substitution codes that are used together. Therefore it is not obvious how to use statistical information on the frequency of the characters in the original text in order to decode the encrypted text, especially when the key is a long word of unknown length. Nevertheless, in the nineteenth century, it was also shown that codes based on polyalphabetic replacement can be attacked using a sophisticated analysis based on frequency analysis of the plaintext character. In particular, the Polish mathematician Kasiski showed how to break the Vigenère code.

6.2.3 The Enigma Machine

Subsequently other secret codes were developed using techniques other than polyalphabetic replacement. This includes the code implemented in the Enigma machine, introduced initially for commercial purposes in Germany from the early 1920s.

The Enigma machine was based on a sophisticated method of substitution (similar to Vigenère's code) that allowed one to define an extremely large number of substitutions. Other methods were also used to make it more resilient to attacks by cryptanalysts. The encoding and decoding functions were very complicated and prone to errors if done by humans; for this reason these operations were done by a machine (the Enigma machine) similar to a typewriter. It was modified and extended over time with new features and subsequently adopted by the Axis forces during World War II.

The Enigma code was "broken" during the war with a method that allowed one to decode a percentage of intercepted secret messages in a relatively short time. In particular, a key contribution to the initial breaking of the Enigma code was made by Polish mathematician Marian Adam Rejewski, who in 1932 was able to identify weaknesses in the algorithm text transformation function of the commercial type of machine then in use. A few weeks before the start of World War II, Polish intelligence services informed the British and French agencies of what they had discovered. Subsequently Rejewski's ideas were used by British intelligence to attack the machine used by enemy forces in the intelligence center at Bletchley Park. Alan Turing (see Chap. 1) had a key role in the design of the attack that was able to decipher in a relatively short time a considerable part of the intercepted encrypted messages.

The influence of this advantage for Allied forces in the war is difficult to assess. It is believed that breaking the Enigma code shortened the war by at least a year.

6.3 Cryptographic Codes and Secret Keys

Cryptographic algorithms like the Caesar and Vigenère ciphers transform plain text messages into encrypted messages by replacing one or more characters with other characters. The replacement operation is not necessarily the same for each character (as in polyalphabetic codes), and in some cases could allow the possibility of adding or removing characters. However, the main operation is to replace a sequence of characters by another sequence.

We also discussed the distinction between the cryptographic code (or algorithm) and the key used by the method. For example, with reference to the Caesar cipher, if we use the English alphabet, we can translate to a number of positions between 1 and 25 and we have 25 possible keys (in fact, if we translate the text of 26 characters we get the original text). The Caesar cipher is not unique in choosing a numeric key: all codes that are in use today depend on alphanumeric keys.

The main difference with respect to the past is that most modern cryptographic algorithms exploit mathematics, and for this reason we can assume that the plaintext message being encrypted is not a literary text but a sequence of integers. In this way, a cryptographic algorithm can be considered as a mathematical function that, given a sequence of integers and a key, calculates a new sequence of integers. The use of mathematics allows cryptography to acquire a sound scientific basis by exploiting the tools and the knowledge developed in the past; moreover this approach also allows the use of mathematics to assess the scientific strength of an encryption code and to verify how to cope with cryptanalytic attacks.

Namely, from the mathematical point of view, the text transformation operated by a cryptographic algorithm C is a function f_C of two integers, the number M which encodes the message, and the key x, and produces a new number Y representing the encrypted message. Clearly a change in only one of the two parameters changes the result. Formally we can write

$$f_C(M, x) = Y.$$

We have already seen the benefits of being able to use the same encryption method to exchange confidential messages with many people; we just need to be careful to define different keys for each communication. As an example, if we encrypt the same message with two different keys x and z then $f_C(M, x)$ and $f_C(M, z)$ will represent the two encrypted messages (that, in general, will be different).

6.3.1 How to Encode a Long Message Using an Integer Function

It is natural to wonder if being able to encrypt numbers is not a limitation, given that the messages that we want to encode are usually text messages. For this reason, we see now how it is possible to transform a text message into a sequence of numbers.

To this goal, suppose we need to encode a text of alphabetic characters and spaces using a sequence of integers. If we use the English alphabet, one possible method is to match the sequence 00 to the blank character, the sequence 01 to the letter A, 02 to B, and so on, by finally matching 26 to Z. Using this correspondence between characters and numbers to assign a number to a text, we can simply concatenate the numbers representing the individual characters, for example, "HELLO" corresponds to the number 0805121215.

Note that the method can also be applied to the case where the text is not only composed of alphabetic characters and spaces, but is a complex text that includes other characters such as upper- and lowercase alphabetic characters, numerals, punctuation characters, and characters that are present in the alphabet of other languages. It is clear that to apply the method we need to use large enough numbers to encode each symbol of interest. We also remark that the method is also applicable to the encoding of other information sources such as music, video, and images: in each case we can consider the message to be encoded as a sequence of numbers.

The above method has the disadvantage that if the message is very long then the number that represents it is very large. For example, to encode a message consisting of 100 characters and spaces with the above-presented method, we need an integer number of 200 digits. Better encodings are used than the simple one we discussed above; however, the number of digits grows linearly with the length of the message. Therefore, the encoding of very long messages requires us to work with very large numbers that not even the computers of today are able to quickly elaborate.

Let us now see how it is possible to encode very long messages using not a single number, but instead a sequence of numbers, each of which has a predetermined number of digits. For example, we can encode a blank space "00", the letter A with "01", B with "02", an so on. In this way, to encode the first verse of Shakespeare's Macbeth ("when shall we three meet again") using a sequence of numbers, each of ten decimal places, we can divide the text into blocks of five characters and spaces and then encode each block with the above-proposed method. Namely, the first block encodes the first word "when" followed by a blank space, the second block encodes the word "shall" and so on. In this way we can represent the verse "when shall we three meet again" composed of twenty-five alphabetic characters and five spaces with the following sequence of six numbers composed of ten digits each

2308051400 1908011212 0023050020 0818050500 1305052000 0107010914

In this way, an encryption method that encodes only ten-digit integers allows us to encode very long texts by dividing them into blocks and encrypting each block separately. Without going into more detail, we observe that the cryptographic methods in use today also divide the message into blocks and then encode the blocks, one at a time, in a slightly more complex way to increase security (see [78, 104]). Furthermore, we note that the length of the block that can be encoded with a number varies significantly depending on the encryption method: some methods have a predetermined length; others admit a variable length.

6.3.2 Cryptanalysis and Robustness of a Cryptographic Protocol

We already observed that the design of cryptographic algorithms and the search for their weaknesses can be seen as a continuous struggle between the inventors of secret codes and the codebreakers whose goal is to "break" the secret code, i.e., being able to decode encrypted messages in a relatively short time. To better understand the rules of this confrontation, we need to define what is meant by secure coding.

We will see that such a definition cannot be given in a few words and requires a thorough discussion. A first approach might claim that an encryption method is safe if the opponent (the cryptanalyst) that reads an encrypted message Y obtained by encoding M is not able to have any partial information on the message unless he knows the secret key used for the encoding of M.

The above statement, if interpreted literally, requires that for the intruder it is impossible to have any information about the message; this safety requirement, also called perfect security, can be obtained, but only with methods that require very restrictive conditions that are not practical. Although the cryptographic methods that have been proposed and are used are not perfect, they nevertheless allow us to achieve security adequate for all practical purposes.

We will see that the conditions that an encryption method must verify to be considered safe are several. The first one concerns what the cryptanalyst knows. In fact, in order to increase the trust in a protocol, we assume that the opponent knows the cryptographic algorithm but also knows the encryption of a set of plaintext messages.

> **Condition 1 – What the Cryptanalyst Knows**
>
> The cryptanalyst knows the encryption method, but he does not know the key used. Furthermore, the cryptanalyst knows a list of encrypted messages and the corresponding plaintexts.

Namely, if $\langle M_1, Y_1 \rangle$, $\langle M_2, Y_2 \rangle$, ... are the pairs of messages and the corresponding encoded messages known to the cryptanalyst, we want that, even with this information, our opponent cannot determine the message that matches to a ciphertext Y (which is different from those already known).

Obviously, knowledge of the method allows the cryptanalyst to know the message that corresponds to Y when he knows the key used for encoding. We also observe that knowledge of the encryption method used also provides partial information on the key. For example, knowledge that the Caesar cipher was used informs us that the key is a number between 1 and 25 (if we use the English alphabet). Similarly, in other cases knowledge of the method provides us with information concerning the set of possible keys; for example, we might know that the key is a number of 13 decimal digits or the key is a prime number of 200 decimal digits.

Information about the encryption method and the type of key used provide the opponent with a chance to decode the message Y. For example, in a Caesar cipher there are only 25 possible keys and therefore in a short time (even without the help of a computer) the adversary that knows a single pair $\langle M, Y \rangle$ of a message and the corresponding encoded message can easily find the key by testing all possible 25 keys and verifying which one transforms M into Y. Note that the same approach can be also used if the cryptanalyst only knows the encoded message Y.

This attack, also called the "brute force" method of analysis, does not use encryption but merely enumerates the possible keys and is always applicable. Therefore, must we conclude that there are no safe encryption methods when Condition 1 is verified?

The answer is not a simple yes or no; in fact, enumerating and testing all possible keys is always possible in theory but often not feasible in practice. Indeed, suppose that the adversary knows that the key is a number of 13 decimal digits. Also in this case the adversary can try all possible keys and surely, sooner or later, will find the right key, but in this case, the possible number of keys is about 10,000,000,000,000. Assuming that the opponent is able to verify a key in a millionth of a second, it takes about 2 months to try only half the possible keys.

The above discussion shows a direct correspondence between computing power used by the opponent and the time to try all possible keys. For example, if the opponent uses 1,000 computers and programs them carefully so that each verifies different keys, then in a relatively short time he may find the right key. But if the key is a number of 200 decimal digits, then using thousands of computers to try only half the keys will still take thousands of years. We are now able to better define the conditions on the methods that can be considered safe on the basis of time and computing resources necessary for the opponent to try all possible keys.

> **Condition 2 – Number of Possible Keys**
>
> The number of possible keys must be very large, in order to prevent the opponents who know the method but who do not know the key from being able to try all possible keys in a limited time.

The former condition requires some clarification and comments. We first observe that the definition of time is deliberately vague: the condition does not define exactly the number of possible keys. In fact, the evolution of computer technology allows for ever-faster computers, so the number of possible keys that can occur in a given amount of time increases year over year. For example, codes with secret keys of 18 decimal digits were considered safe 30 years ago, but now new codes are being proposed based on keys with at least 40 decimal integers; in the future we will use codes with even longer keys.

The second observation is that there is no absolute concept of security, but it is a concept related to the specific field of application. Indeed, the secrecy of a message related to the purchase of an object in an electronic auction is a condition which must hold for the time duration of the auction (usually no more than a few days), while the confidentiality of a message related to the national security of a country must remain secret for a very long time, even tens of years. Therefore, the concept of the limited time of Condition 2 also depends on the particular application and on the message.

We also observe that Condition 2 does not exclude the possibility of a lucky adversary guessing the key with a few attempts. In fact, if the opponent is very lucky he can also try the right key on the first attempt! However note that, if the number of keys is very large, the probability that this occurs is very low, almost negligible.[4]

The above discussion allows us to conclude that Condition 2 is a necessary condition for the design of a secure code. However, this condition is not sufficient. The ultimate goal of a cryptanalyst is to be able to identify the key used to encode

---

[4] As an example, the probability of guessing a secret key of 40 decimal integers is smaller than the probability of having 130 consecutive heads while tossing a fair coin.

the text with a limited number of attempts. For example, if an encryption code includes a set of possible keys, one billion of billions and billions of keys can be considered safe enough, but if a cryptanalyst is able to reduce the number of possible keys to only one million then we can assume it is an insecure code: a computer is able in a short time to try a million keys and find the right one. In fact, there are codes which, while having a very large number of possible keys, have intrinsic weaknesses that allow the cryptanalyst to identify their keys with a relatively small computational effort. Examples of these weaknesses are the statistical analyses that have been discussed previously for codes based on replacement, and that show how the definition of a secure code cannot ignore the skills and ingenuity of cryptanalysts.

Recall that Condition 1 assumes the cryptanalyst might know several pairs ⟨Messages, Encoded Message⟩; moreover, when trying to decrypt an encoded message Y, the cryptanalyst might also have partial information about the message that generated Y. For example, in a military setting, if it is clear that an attack of the enemy is about to occur, it is not very important to completely decode the message: the most important thing in this case is to discover the time and exact location of an attack.

We are now ready to summarize the discussion on when an encryption code should be considered secure: an adversary who knows the method and a list of unencrypted and encrypted messages should not be able to deduce information about a message encrypted with a key different from those known to him.

Condition 3 – When an Encryption Code Is Safe

An encryption code is to be considered safe when, given an encoded message Y, the opponent is not able in a reasonable time and with non-negligible probability to obtain any partial information on the plaintext M corresponding to Y or to find the secret key that was used to encode Y.

This should not be possible even if the adversary knows the encryption method that was used, a list L of messages and their corresponding encrypted messages (assuming that L does not contain M), and he cleverly applies known cryptanalysis methods.

The previous condition summarizes all the requirements that an encryption code must satisfy. From a mathematical point of view, the above condition means that the knowledge of $Y = f_C(M, x)$ must not yield information on M or on x to an opponent. This condition should occur even when the opponent knows the function f_C (in fact, assume that the adversary knows the encryption method) and knows the value of the function for a certain number of values (the messages of the list L).

6.4 Secret Key Cryptography

Cryptography algorithms that we discussed in Sect. 6.2 assume that before exchanging confidential messages, the sender and the recipient must choose a cryptographic algorithm *and* a secret key; to preserve the confidentiality of the communication the chosen key should be protected and not revealed to anybody else. This type of encryption, known as secret key cryptography or symmetric cryptography, was the only type of encryption known until 1976 and it is still largely used.

Nowadays, many secret key cryptographic codes that are used can be classified as block codes; a block code can be considered a modern version of the polyalphabetic substitution code proposed by Alberti: the code receives as input a sequence of characters of a fixed length and a key, and outputs a block of characters with the same length. The length of the input block varies according to the cryptographic method, but in most cases is less than 200 bits, thus allowing us to encode only a fragment of the message. Since messages are almost always longer than this, the text block to be encrypted in general consists of several blocks and it is therefore necessary to use a method to encode a sequence of blocks. We saw such a method in Sect. 6.3.1; we do not enter into a discussion of the security of such a mode, which goes beyond the purpose of this chapter. We simply observe that particular care must be used while encoding long messages using a block code so that no partial information is disclosed to the adversary.

6.4.1 Secret Key Cryptography Standards

From 1960 onwards, electronic computers have become increasingly powerful and economical, and have become increasingly popular in business, and about 20 years later they entered private homes. The parallel development of telecommunications networks has opened up new scenarios for the use of encryption and has posed new problems. The first one is to determine a standard algorithm, i.e., an algorithm that is known worldwide and is recognized as being secure.

To meet this need, in 1976 the US National Standards Agency (NSA) adopted the Data Encryption Standard (DES) as the secret key encryption standard; DES is a block code with 56-bit keys. Today, DES is no longer the standard because it is not adequate to current security needs. In fact, the number of different keys that an attacker must try is limited with respect to the computing capability of today's computers; therefore, nowadays an attack on DES can be performed in a relatively short period. Moreover, 56-bit DES has also been the subject of cryptographic attacks that significantly reduce the number of keys that an attacker must try to find the key to a coded message to about eight trillion (a large number but not too large to perform an exhaustive search in a relatively short time with the computers available today).

Today the standard for secret key encryption is the Advanced Encryption Standard (AES), proposed by Rijmen and Daemen. In its basic version AES allows one to encode a block of fixed length of 128 bits using keys of 128 bits. The AES encryption is performed by repeating a "round", which is a sequence of operations that transforms a block into another block using the secret key. When all rounds are completed, the block obtained represents the ciphertext of the message. The same operations are performed in reverse order to get the initial text from ciphertext.

Regarding the security of AES, we first observe that the number of possible keys that can be defined using 128 bits is very large, so large that an exhaustive search of all keys cannot be performed in a century even if an opponent uses a million computers that are the fastest on the market today. Furthermore, none of the known cryptanalysis attacks can significantly reduce the number of possible keys. Thus, AES is now considered to be robust, but this excludes any weaknesses that may be found in the future. If these weaknesses would call into question its safety then inevitably we would need to define a new standard.

6.4.2 Limitations of Secret Key Encryption

The definition of a reference standard for secret key cryptography, which occurred in 1976 with DES, has allowed for extensive development of civil uses of cryptography, first in finance and then in other economic sectors. At the same time, the widespread utilization of secret key cryptography has shown two main limitations of this approach to cryptography that apply to all such algorithms. In the following we discuss these limitations and the solutions that have been proposed.

The first limit of secret key encryption is the requirement that, before starting the secret exchange of messages, it is necessary to agree on the secret key that will be used. In many cases, such a requirement is a significant limitation to the use of cryptography. As an example, imagine an Italian company which wants to start a business relationship with a Chinese one and for this reason wants to send confidential messages using e-mail. To be on the safe side, the company wants to use encryption in order to exclude the possibility that a competitor is able to read exchanged e-mails.

To this aim, the company first proposes to start a confidential communication using a secret key algorithm; however, this requires the creation of a secret key between the two partners. We can assume that one of the two partners proposes the secret key, but then how can the other partner know the key? Clearly it is unacceptable to first exchange the secret key using a nonencrypted e-mail message; in fact, the initial message containing the key might be intercepted and read by an adversary. The problem looks like the "chicken or the egg" causality dilemma: we should be able to encrypt the initial message that contains the secret key, but to do this we must have a secret key.

For several years the only solution to the above problem was an expensive one: exploit trustworthy people who physically transferred the keys. This solution is not only expensive but also not completely secure (because it relies on the physical exchange of the secret key).

The second problem of secret-key encryption is that it requires a key for each pair of communications. In particular, if a user wants to communicate secretly with 100 different friends, then 100 different keys are required. In general, if many people want to communicate with each other in confidence, then the number of keys required increases greatly. In fact, 100 people who want to communicate in pairs using different keys must use 4,950 different keys; you can verify that if n is the number of people who want to communicate with each other, then the total number of keys required is $n(n-1)/2$.

The increased number of keys that are required is a limitation on the spread of encryption because it poses problems concerning the secure storage of large sets of keys. A detailed discussion of such issues goes beyond the purposes of this chapter; we simply observe that it is commonsense that the secret keys should be safely stored to not allow intruders to obtain information. Therefore, while we are able to memorize small numbers (like we do with the PIN of our ATM card), it is almost impossible to be able to memorize hundreds of different keys, each one of many 128-bits like AES secret keys.

6.5 The Key Distribution Problem

We discussed in the previous section that secret key cryptography requires parties to agree on the secret key to be used before establishing the confidential communication. Ideally, two users could decide the secret key without meeting, but using a communication line that is not secure and that can be tapped by an adversary.

Defining a secret key between two users using a nonsecure communication channel is a challenging problem that at first sight looks impossible. In fact, we require that even though the adversary is able to listen to all exchanged messages, he should not be able to know the secret key. Surprisingly, in 1976 Diffie and Hellman proposed a simple solution to the problem. Their method is still used today and represents one of the most important discoveries in the history of cryptography, not only for the relevance of the problem but also for the introduction of innovative algorithmic and mathematical techniques that paved the way to a new approach to cryptography. In particular, Diffie and Hellman proposed the use of mathematical functions that are easy to compute but difficult to invert.

Many mathematical functions are both easy to compute and easy to invert. Here "easy" and "difficult" should be intended with respect to the computation time required by a solution algorithm; namely a problem is easy if there is a polynomial-time algorithm that solves it, and it is hard if such an algorithm cannot exist or it is believed that it does not exist (see Chap. 3). For example, the function that computes the double of a number is easy to compute and it is also easy to reverse: the calculation of the inverse of the function "double" is half of a number. Since such a calculation can be done even for very large numbers we can therefore say that the function "double" is easy to calculate and easy to reverse.

Functions that are easy to calculate but that are difficult to invert play a fundamental role in cryptography. Because of their special features they are also called "one way". Diffie and Hellman's proposal exploits one such function; before presenting their method we must introduce some concepts of modular arithmetic.

6.5.1 Modular Arithmetic

While in the usual arithmetic with integers we perform operations using the set of all integers, in modular arithmetic we restrict ourselves to a restricted set of possible values. For example, in arithmetic modulo 5, possible values are only 0, 1, 2, 3 and 4; similarly, in arithmetic modulo 12 we use integers between 0 and 11.

The rules by which we carry out the operations in modular arithmetic take into account the limited number of values available. In particular, if the calculated value is less than or equal to the maximum representable number, then the result is the same as in the usual arithmetic. For example, the addition of 2 and 2 in arithmetic modulo 5 (which is written $(2 + 2)$ mod 5) gives 4. Instead, if the result in the usual arithmetic is a value that cannot be represented, the calculation is slightly more complex. For example, to calculate $(2 + 6)$ mod 5 (the sum of 2 and 6 modulo 5) we can imagine arranging the numbers between 0 and 4 on a circle, as on the face of a clock, with the numbers indicating the hours. At this point, to calculate $(2 + 6)$ mod 5 we start at 2 and move clockwise for 6 positions: after making a complete revolution of the circle we get to number 3, which is the result of $(2 + 6)$ mod 5.

At first glance this approach might seem complicated, but note that we do this type of operation when we calculate the hours of the day. For example, if it is 9 a.m. and we have an appointment in 6 h we can say that our meeting will be at 3 p.m. In this case we compute $(9 + 6)$ mod 12, which is 3. We observe that, from a mathematical point of view, the calculation of $(9 + 6)$ mod 12 is equivalent to first computing the sum in the usual arithmetic (obtaining 15), and then by considering the remainder of the division of 15 by 12. Therefore, on the basis of what has been discussed, we observe that $(3 + 7)$ mod $5 = 0$ and that $(11 + 5)$ mod $10 = 6$.

Similarly, if we perform the multiplication operation in modular arithmetic we can proceed in two steps: first computing the product of the two numbers in the usual arithmetic and then computing the remainder of the division of the answer obtained divided by the module. For example, suppose we have to calculate $(2 \cdot 6)$ mod 5; we first compute in the usual arithmetic $2 \cdot 6$ (which is 12), and then we compute the remainder of the division of 12 divided by 5. Therefore 2 is the result of $(2 \cdot 6)$ mod 5. Similarly, $(3 \cdot 5)$ mod $7 = 1$; in fact 1 is the remainder of 15 (which is $3 \cdot 5$) divided by 7.

In modular arithmetic both addition and multiplication are easy to calculate (i.e., computable in time polynomial in the length of the input) and also their inverse functions are easy; for this reason addition and multiplication have a limited interest in cryptography. An operation of particular interest is instead modular exponentiation. Observe that the calculation of powers in modular arithmetic is an easy operation as in the case of multiplication: for example, if we calculate 2^3 mod 5 we first compute $2^3 = 8$ in the usual arithmetic and then perform the remainder of the division of 8 by 5. We then have 2^3 mod $5 = 3$.

Exponentiation in modular arithmetic preserves many of the properties of exponentiation in arithmetic but has a special feature that makes it very useful in cryptography. In fact, the calculation of the remainder of the division implies that the function of exponentiation behaves in an apparently bizarre way. For example, suppose we calculate x^2, the square of an integer x. In the usual arithmetic the square of a number is a function whose value increases as x increases. In fact, if $x = 3$ then we get $3^2 = 9$ and if x is equal to 5 we get 25. This observation facilitates the calculation of the square root, the inverse of the square function; namely, given an integer z suppose we need to compute y such that $y^2 = z$. In particular, given $z = 16$ and knowing that $3^2 = 9$ and that $5^2 = 25$, by observing that 16 is greater than 9 and smaller than 25 suggests that the square root of 16 must be between 3 and 5 (indeed, 4 is the square root of 16).

If we use modular arithmetic the above feature vanishes; in fact, if we perform the same operations modulo 11 we get (3^2) mod $11 = 9$ and (5^2) mod $11 = 3$; namely, we have that the square of 3 is greater than the square of 5. This does not help us to calculate the square root of 5 in arithmetic modulo 11, which is 4 (in fact (4^2) mod $11 = 5$).

The above example illustrates an operation for which the calculation of the inverse function is actually more complicated. It is important to underline that we do not know any fast algorithm to compute the inverse of the exponentiation function in modular arithmetic when operations are carried on modulo n, when n is a very large prime number, for example, composed of hundreds digits.[5]

6.5.2 Diffie and Hellman's Algorithm for Establishing a Secret Key

Suppose that two friends, Alice and Bob, need to define a secret key using an insecure communication line; namely, we assume that an intruder, which we denote as Charles, is able to tap all exchanged messages between Alice and Bob. The method proposed by Diffie and Hellman for establishing a secret key under such conditions, requires that both Alice and Bob execute the algorithm independently of each other and exchange data that is the result of random choices that are not revealed. An algorithm of this type is generally presented in the technical literature as a protocol because two independent parties must execute it.

---

[5]In fact, even though we do not have a formal proof of intractability, the problem is considered intractable.

The basic idea is to use a function of the form $(y^x) \bmod z$. At the beginning, Alice and Bob will choose two integers y and z. The definition of these two numbers can be made in the clear (and thus allowing Charles, the intruder, to know both y and z) with a few restrictions: z must be a prime number (i.e., a number that has no divisors other than 1 and z itself), and y must be smaller than z. Once y and z have been chosen, the protocol requires that both Alice and Bob each choose a secret number; assume that a and b are the two numbers chosen by Alice and Bob, respectively. For the security of the protocol it is important that both a and b are not disclosed but remain secret.

The secret key s that Alice and Bob will use is given by the $(a \cdot b)$th power of $y \bmod z$; formally we have

$$s = y^{a \cdot b} \bmod z.$$

Note that Alice does not know b (and symmetrically Bob does not know a). Therefore to allow both parties to compute s the protocol continues with the exchange of two messages: Alice sends $(y^a) \bmod z$ to Bob; Bob in turn sends to Alice $(y^b) \bmod z$.

At this point, Alice and Bob have all the information to calculate the secret key s. In fact,

- Alice knows y, z, a and $(y^b) \bmod z$. To compute s she performs the ath power of what she received from Bob; formally Alice performs the following calculation

$$[(y^b) \bmod z]^a \bmod z.$$

- Similarly, Bob computes the bth power of what he received from Alice obtaining

$$[(y)^a \bmod z]^b \bmod z.$$

The modular arithmetic properties imply that the two numbers are equal. In fact, the following equality holds

$$[(y^b) \bmod z]^a \bmod z = [(y^a) \bmod z]^b \bmod z = s.$$

In this way Alice and Bob have defined a secret key by exchanging information relating to secret numbers a and b but without exchanging the numbers themselves.

We now analyze the security of the method: we assume that the intruder is able to know all exchanged messages between Alice, and we wonder whether he is able to reconstruct the secret key s. The information available to Charles are the initial numbers y and z and the two exchanged values $(y^a) \bmod z$ and $(y^b) \bmod z$; however he does not know numbers a and b, which are kept secret. This is the key point of the algorithm: knowledge of $(y^a) \bmod z$, y and z implies that to compute a Charles must be able to perform the inverse operation of exponentiation modulo z. As we discussed in the previous section, the computation of the inverse of exponentiation in modular arithmetic is a function that nobody knows how to compute in polynomial time, and therefore it is impossible when the module is a large prime number.

We observe, however, that Charles has the ability to check whether a number is indeed the secret number chosen by Alice (or by Bob). However, since a and b are chosen at random between 1 and z, it follows that the probability that Charles' guess is correct can be made arbitrarily small by choosing a very large value of z. For example, the choice of z may be sufficiently large so that the probability of guessing a or b is negligible; for example, it can be less than the probability of guessing for 10 consecutive times the winning numbers of the lotto! In conclusion we have shown how Alice and Bob were able to calculate a secret key between them with a robust method with respect to an attacker who is able to listen to their communication.[6]

6.6 Public-Key Cryptography

Public-key cryptography is a radical discovery useful for establishing confidential communication and that allows us to obtain new applications of cryptography. The idea of secret key cryptography is to associate a secret key to each pair of people, thus allowing them to establish a confidential communication. Differently, in public-key cryptography keys are associated to each *single user*; namely, for each user we have a *public key* that can be used to encrypt messages and a *secret key* that can be used to decrypt messages encrypted. The encoding is such that decryption requires knowledge of the secret key.

To better illustrate the concept we use a metaphor in which a secure box is used to transmit the message; the box can be closed with a lock by the sender of the message and then it is sent to the recipient. During travel the secure box cannot be opened; upon arrival the recipient of the message opens the lock and reads the message. To represent the above process using a secret-key encryption method we can imagine that Alice and Bob buy a lock and each of them makes a copy of the key. In this way they can safely communicate: when Alice wants to send a message to Bob, she puts it in the box and locks the box with the lock. When Bob receives the box he opens the lock using his key and gets the message. Clearly, the communication from Alice to Bob is analogous and can be done using the same key lock. Obviously, the accuracy of the method is based on the fact that only Alice and Bob own copies of

---

[6]The adversary who only listens to the communication is called a passive attacker and is the weakest possible adversary. An active adversary is one who is able to send messages on behalf of others and possibly intercept and modify messages posted by others. The method proposed by Diffie and Hellman is secure with respect to a passive adversary but it is not resilient with respect to an active attacker. We refer to [66, 78, 104] for how to cope with active adversaries.

the key; therefore, if an intruder had a copy of the key he would be able to open the box and read the messages in it.

The above scenario describes a system characterized by a perfect symmetry between Alice and Bob, and where the key represents the secret encryption key.

In the case of public-key cryptography, the transmission of messages to Alice using the locked box can be modeled by imaging that Alice buys many copies of the same lock. Namely, we assume that all locks can be opened by the same key and that to close an open lock we do not need the key. Alice distributes the open locks to anyone who wants to send her a message and holds the keys to the lock itself. When Bob want to send a message to Alice he puts it in the box, closes the box with the lock that Alice gave him, and sends the box back to Alice. Upon receiving the box Alice opens the lock with her key and gets the message. The crucial observation is that only Alice holds a copy of the key, and therefore she is the only one able to open a closed box. The above scenario models a public-key cryptography method.

We observe an asymmetry between Alice and Bob: Bob has a copy of the open lock, supplied by Alice, but he is unable to open the lock, which may be done only by Alice. In particular, the open lock represents the public key, and the key represents Alice's secret key.

The idea of public-key cryptography was presented to the scientific community by Diffie and Hellman in an article published in 1976, without proposing an effective public-key cryptographic algorithm, with appropriate security. Their paper, however, represents a milestone in the history of cryptography, because it proposes a revolutionary approach, completely different from the methods used in more than 2,000 years of history. The year after the publication of the article by Diffie and Hellman, Rivest, Shamir and Adleman proposed the RSA encryption algorithm (named from the initials of the authors). The salient feature of the method is that until now nobody has been able to find a possible attack that makes it unusable; for this reason today RSA public encryption is the most widely used standard.

From the historical point of view we now know that Diffie and Hellman were not the first to have had the idea of public-key cryptography: Cocks and Ellis, two English mathematicians who worked for the British intelligence services proposed the idea of a public-key cryptosystem scheme together with a cryptographic algorithm—similar to the RSA algorithm—in an internal document in 1973. The revolutionary contribution of their proposal was misunderstood by their bosses (perhaps because of the slowness of computers needed for its implementation at the time) and considered a curiosity. However, the proposal was classified top-secret and for this reason it was not made public until December 1997.

6.6.1 The RSA Algorithm

We proceed in two steps to present the RSA encryption algorithm; first we will see how to define the pair of keys, the public key and the secret one, that will be used by Alice, and then we will see the encoding and decoding algorithms.

6.6.1.1 How to Define the Public and the Secret Keys

Assume that Alice wants to establish a public key and corresponding secret key. To this goal she first chooses two large prime numbers p and q. For security reasons p and q must be large—they are composed of hundreds of digits—and it is essential that they are kept secret since knowledge of them would allow an attacker to easily break the method by computing the secret key. We do not discuss an efficient procedure for finding such primes, and instead we refer to [66, 78, 104].

Alice's public key is given by a pair of integers (n, e), where n is the product of p and q, and e is an integer greater than 1 that is relatively prime with $(p-1)\cdot(q-1)$.[7] To find such an e is not too complicated: Alice chooses a random integer repeatedly until she finds one that satisfies the required condition; in general this procedure does not take long because there are many numbers that have the property of being relatively prime with $(p-1)\cdot(q-1)$. This step completes the definition of Alice's public key.

Alice's secret key that is associated with the public key (n, e) is defined by the integer d such that $e \cdot d = 1 \bmod ((p-1)(q-1))$. Note that given (n, e), then d is unique and its calculation is not computationally difficult as it can be done with a variant of Euclid's algorithm for calculating the greatest common divisor of two numbers (see Sect. 1.3).

6.6.1.2 How to Encode and Decode

Recall that in Sect. 6.3 we showed how we can represent messages using integers; in particular, we assume that the message m is an integer less than n. The RSA encryption algorithm takes as input the public key (n, e) and m and computes an integer y, which is also less than m, representing the encoded message; the secret key d allows us to compute m from y.

The encoding operation simply requires us to calculate the eth power of m modulo n. Formally, we have

$$y = m^e \bmod n.$$

Decoding is similar: to obtain m it is sufficient to calculate the dth power of y modulo n:

$$y = m^d \bmod n.$$

---

[7] Recall that e and $(p-1)\cdot(q-1)$ are relatively prime if the Greatest Common Divisor of e and $[(p-1)\cdot(q-1)]$ is 1.

To show that the RSA method is correct we need to show that, given m by first encoding and then decoding we get the original value m. Without going into the details of a rigorous treatment, we simply observe that the definition of the different parameters n, e and d implies that, given m, first computing the eth power of m modulo n and then the dth power of what it achieved (always modulo n) allows us to get m again. Equivalently we can claim that the encoding and the decoding operations are the inverse of each other, and therefore we can write:

$$m = y^d \bmod n = (m^e \bmod n)^d \bmod n.$$

6.6.1.3 Security of the RSA Algorithm

We first observe that Alice's public key (n, e) is known by everybody, even to the adversary who wants to know the confidential messages sent to Alice. Therefore, the security of the method hinges on the fact that knowledge of the public key does not allow one to know the associated secret key. Recall that the secret key is uniquely associated with the public key; therefore the security of RSA requires that the computation of the secret key, knowing the public key, should be computationally intractable.

Observe that if an attacker knows the public key (n, e) and knows the factors p and q of n, then he is able to calculate the secret key. In fact, the knowledge of p and q allows one to calculate $(p-1) \cdot (q-1)$ and, subsequently, knowing $(p-1) \cdot (q-1)$ and e it is very easy to compute the secret key d by performing the same operations that Alice has done while generating her keys. Therefore, the safety of RSA hinges on the fact that given an integer n product of two primes, the computation of the primes p and q such that $n = p \cdot q$ is computationally hard. This problem is a special case of the integer factoring problem that asks for the prime factors of a given integer.[8]

Finding an algorithm that solves the factoring problem is not difficult: there is a simple resolution algorithm that is based on repeated checking of whether a prime number is one of the factors. Namely, to find the factors of n it is sufficient to divide it by 2 and verify if the remainder of the division is 0. In the positive case we have found that 2 is a factor of n; if not we carry out the division of n by 3 and verify the remainder. If also in this case the remainder is not zero then we continue with 5, then 7 until we find a factor or we reach a sufficiently large number (greater than the square root of n) without finding the prime factors of n. In this second case we can deduce that n is a prime number and as such has no prime factors other than 1 and itself.

---

[8]Recall that f is a prime factor of n if 0 is the remainder of the division of n by f and f is different from 1 and n (in fact, we can also say that 1 and n are trivial factors).

The previous algorithm is conceptually simple but impractical if n is very large. For example, the factorization of a number given by the product of two primes p and q of 100 decimal digits each (whose product n is composed of 200 decimal digits) requires the execution of a very high number of divisions, which makes the method impractical. We might wonder whether there are more efficient algorithms that make the factoring problem tractable. The answer is negative: despite the intensive studies done we do not know fast methods that are capable of factoring numbers that are the product of two large prime numbers. In fact, the fastest methods proposed are able to factor numbers of 200 decimal digits (in a few months and with the use of parallel computers) but are impractical to factor larger numbers. In fact, the computation time of all known algorithms grows exponentially with the number of digits of the number to be factored if both p and q are very large. In particular, known algorithms may require centuries to factor integers of 400 decimal digits (such as those used today in RSA applications), even if we have thousands of available computers.

The above discussion shows that with the state of the art we do not know whether a fast algorithm to factor an integer that is the product of two prime numbers exists or not. If such an algorithm is discovered, then the RSA encryption code is no longer secure; however, since nobody has been able to find such an algorithm despite the problem having been intensively studied is a sufficient reason to believe that *for now* RSA is secure. We have emphasized now, because the previous claim does not exclude the possibility that a brilliant researcher might have the right approach to solve the problem. In other words, we have no certainty of the fact that RSA is secure, but there no algorithms that allow an attacker to infer the secret key from the public key. We can only say that RSA should be considered secure given the present state of our knowledge.

6.7 Digital Signatures and Other Useful Applications of Public-Key Cryptography

In the last 20 years we have witnessed a radical change in the way information is stored and distributed; during this time paper has been progressively substituted by digital media. This radical change poses new problems: for example, when we receive a file, how can we be sure that such information is authentic and has not been modified?

The above problem is caused by a well-known feature of computer and digital devices in use today: the ease with which we can copy and modify information. The fact that copying and modifying are simple poses the question of how to guarantee data integrity and authenticity. Therefore, confidential communication is not sufficient to meet all safety requirements posed by the many applications of new information technologies such as e-mail, and Internet commerce. In particular, since information is primarily stored and transmitted in electronic format, it is necessary to develop methods and tools to ensure the authenticity of the information and of the sender.

The issue of authentication is not limited to data. In fact, the original design of the Internet did not consider security and privacy aspects; therefore the original design does not guarantee the identity of the person with whom we are communicating (as we often experience with fake e-mails that claim to be sent by our bank and ask to verify the access codes of bank or credit card accounts).

The above issues have many consequences that are out of the scope of this chapter such as, for example, that of copyright protection. For the sake of space we will limit ourselves in the following to discussions of how to digitally sign documents whose importance in our society does not need to be justified.

6.7.1 How Public-Key Cryptography Allows for Digital Signatures

We first observe that the ease of copying a document complicates the problem of signing digitally. In fact, hand signature depends only on the person signing, and might require experts to distinguish fake signatures. In the case of digital signatures the problem is more complicated. In fact, suppose that Alice sends a bank order to Bob for $100 and assume that Alice signs the order by writing "Alice" at the end of the message. Alice's message is a sequence of characters that can be easily copied, so the signature of Alice may be taken by our intruder, say Charlie, who could send to the bank a message like "Pay $1000 to Charlie" and adding "Alice" at the end. Clearly the above example holds, even if Alice's signature is a more complex string such as "A!23jde/Z2".

The definition of a digital signature scheme requires us to define an algorithm for signing a document and an algorithm for verifying whether a signature is a true (or a fake) signature. The above discussion implies that *the digital signature of a document must depend on the signing person and on the document; namely, the signatures of two different documents done by the same person should be different.* However, we now need to define a digital signature method that must allow us to verify the correctness of the signatures created by the same person on different documents.

6.7.1.1 How to Digitally Sign with Public-Key Cryptography

Public-key cryptography allows us to realize in such a way conceptually simple digital signatures. In particular, suppose that Alice wants to send the following message to the bank[9]:

> Pay the sum of 100 US dollars to Bob charging it on my account. Rome, 1.1.2008 8:00.
> Alice

---

[9] The reason for including date and time in the next message will become clear shortly.

6 Algorithms for Secure Communication

Suppose that Alice is using RSA and that (n, e) is Alice's public key and that d is the associated secret key (that is only known to Alice and to nobody else). Let m be the number (less than n) representing the message to be signed. The operations of signing and verifying a digital signature are described as follows.

Signature Algorithm

Alice sends the message m and signs by decoding it with her private key. Namely, the digital signature f is

$$f = (m^d) \bmod n.$$

Signature Verification Algorithm

After receiving the message m and Alice's signature f, to verify the signature it is necessary to know Alice's public key.

Namely, if (n, e) is Alice's public key we define g as follows

$$g = f^e \bmod n.$$

The signature is valid if $g = m$; in such a case the bank executes the payment order, otherwise the signature is false.

To prove the correctness of the above method we must provide evidence that only Alice is able to sign a message with her name while at the same time everyone should be able to verify the signature. Before discussing how this is achieved we recall that when applying RSA encryption and then RSA decryption on a given m then we get m; formally the following equality holds for all m

$$m = (m^e \bmod n)^d \bmod n.$$

The correctness of the verification made is based on the fact that if we perform two successive operations of exponentiation the order is not important. For example, squaring 2 and then computing the cube power of the result gives the same result if we first compute the cube of 2 and then compute the square of the result; in fact we have that $(2^2)^3 = (2^3)^2 = 64$.

Formally, the fact that when executing two successive operations of raising a number to a power we can change the order of execution implies that the operation of exponentiation is a commutative operation. This property is also true in the case when the result of the operation is the module of n.

The above discussion implies that—in the RSA method—given m by first decrypting and then encrypting we obtain the same m as when we first encrypt and then decrypt. Formally, the following equality holds for all m:

$$m = (m^d \bmod n)^e \bmod n = (m^e \bmod n)^d \bmod n.$$

Now recall that the integers n, d and e that form the secret and the public key are strictly related: in particular, given the public key (n, e) there is only one integer d that corresponds to Alice's secret key and verifies the previous equality. Therefore, only Alice, who knows the secret key associated with the public key (n, e), is able to compute, which verifies the equality

$$f = (m^d) \bmod n.$$

On the other hand we observe that to verify the digital signature created by Alice we need to know Alice's public key. However, we can assume that this information is publicly available and therefore everyone is able to verify Alice's signatures. We also observe that any forger of Alice's signature must be able to calculate Alice's secret key from her public key. We have discussed in the previous paragraph that this task is virtually impossible and takes a long time even if you have many computers. Therefore we can safely assume that nobody can use the signature of a message to falsify the signature of other messages.

Note that by applying the above algorithm, Alice's signature depends on the message, and the signatures of different messages are different. Therefore, the previous reasoning to ensure that only Alice is able to sign the message M also implies that Alice cannot repudiate her signature before the judge who asks her to show her secret key. Obviously, Alice's signature of identical messages is the same. This implies that if Alice wants to send two identical payment orders then she should make them different from each other in some detail (such as the date or time of the payment as we did in the example). Otherwise the two messages are as indistinguishable as their signatures.

In conclusion, we can conclude that only Alice is able to sign and, therefore, the validity of the signature itself follows and the proposed digital signature scheme verifies the following requirements.

Requirements of a Digital Signature Algorithm

A digital signature algorithm must be not falsifiable, not repudiable and verifiable. Not falsifiable implies that nobody should be able to sign documents with somebody else's name. The second condition implies that a judge in court should be able to certify the authenticity of the signature (in the same manner as certified by a notary). The third condition requires that everyone should be able to verify the validity of a signature done by somebody else.

Observe that the two requirements of non-falsifiability and non-repudiation are closely linked.

We have seen how to sign a message that is represented by an integer number m. We have seen how we can represent a very long message with a sequence of integer numbers $\langle m_1, m_2, \ldots \rangle$. Therefore, if we want to sign a very long message then it is sufficient to apply the above digital signature scheme to $m_1, m_2, \ldots$, obtaining as a signature a sequence $\langle f_1, f_2, \ldots \rangle$. Observe that this method has a major limitation: the digital signature of a document is as long as the document itself. Clearly this is not practical. We would like to have short signatures. This can be indeed achieved; however, the discussion of this is beyond the purpose of this chapter and we refer to [66, 78, 104].

We conclude by noting that there are other issues in addition to sending confidential information and the digital signature of documents that are processed in cryptography, for example, the integrity of data and identification of users. In the first case you need to have tools to quickly determine whether the information has been modified in an unauthorized way (e.g., by inserting, modifying or deleting data). The identification of users instead mainly applies in scenarios where you need to be certain of your identity.

In the same way as for the sending of confidential information and the digital signature, encryption has developed the modern mathematical methods and techniques that allow you to manage these aspects. We refer to the bibliography for further study.

6.8 Bibliographic Notes

For a history of cryptography, which includes not only the origins but also the latest developments, the reader is referred to [64] and [103]. The first monograph presents a thorough discussion of the development of cryptography over the centuries, and the second monograph is characterized by the historical framework with a series of examples that highlight the role of cryptography.

For a story about the activities of British intelligence during the Second World War which led to breaking the Enigma code, refer to [59]. For more on modern cryptographic codes and methodologies, it should be noted that the encyclopedia Wikipedia contains much useful information. There are many textbooks on the market that present basic elements of modern cryptography, for example, [66, 104]. For a very complete and detailed review of the main topics of cryptography, see the manual [78]; chapters of the manual may be obtained from the publisher's Web site.

Chapter 7
Algorithmics for the Life Sciences

Raffaele Giancarlo

Abstract The life sciences, in particular molecular biology and medicine, have witnessed fundamental progress since the discovery of "the Double Helix". A relevant part of such an incredible advancement in knowledge has been possible thanks to synergies with the mathematical sciences, on the one hand, and computer science, on the other. Here we review some of the most relevant aspects of this cooperation, focusing on contributions given by the design, analysis and engineering of fast algorithms for the life sciences.

7.1 Introduction

In February 2001, the reference journals *Science* and *Nature* published special issues entirely dedicated to the sequencing of the human genome completed independently by *The Human Genome Consortium* and by *Celera Genomics*, with the use of two different sequencing approaches. Those results, of historic relevance, had already been widely anticipated and covered by the media since they are a fundamental landmark for the life sciences—biology and medicine, in particular. Indeed, the availability of the entire sequence of bases composing the human genome has allowed the comprehensive study of complex biological phenomena that would have been impossible before then. The abstraction process that allows a genome to be seen as a textual sequence is summarized in the box "Textual Representation of DNA". The race towards such a goal began in the early 1990s, when it became clear that the sequencing technologies available then, with the focused support of research in mathematics and computer science, could be extended to work on a genomic scale.

R. Giancarlo (✉)
Dipartimento di Matematica ed Informatica, Università di Palermo, Via Archirafi 34, 90123 Palermo, Italy
e-mail: raffaele@math.unipa.it

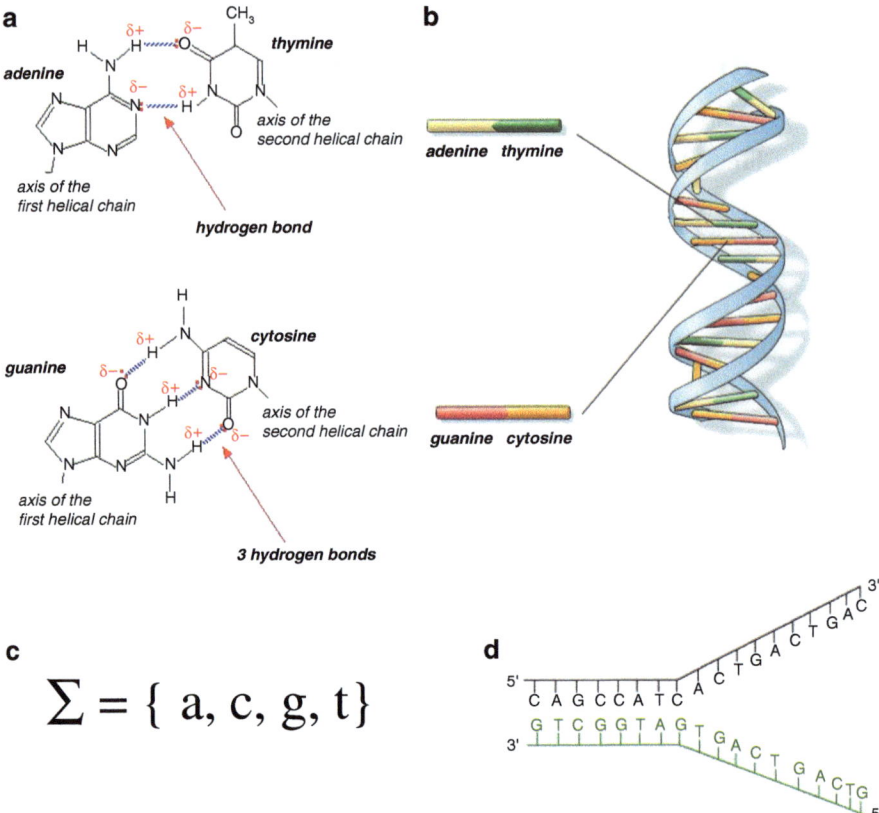

Fig. 7.1 A genome as a text sequence

Textual Representation of DNA

DNA can be represented in textual form, i.e., its biochemical structure can be described by a sequence of characters, as briefly outlined next. The four nucleic acids that compose DNA are Adenine, Cytosine, Guanine, and Thymine. They bind each other in a complementary way, as follows: Adenine and Thymine bind exclusively to each other and so do Cytosine and Guanine. The chemical structure of these bases and their links are illustrated in Fig. 7.1a. The "double helix" is shown in Fig. 7.1b. Its main features are: the skeleton of external support, made from sugars, represented on both sides as a ribbon, and the two filaments of bases linked in a complementary way that are represented as rods. The four nucleic acids can be simply "coded" with letters: A, C, G, T—with obvious association. The complementarity of

(continued)

> **(continued)**
>
> the bonds can be represented by pairs of letters (A, T) and (C, G). Once that is done, the biochemical structures of Fig. 7.1a can be represented by an alphabet of four symbols, shown in Fig. 7.1c. It is worth pointing out that a double-stranded DNA structure can be represented by choosing one of the two sequences corresponding to one of the two strands, since the other sequence can be uniquely determined from the chosen one (see Fig. 7.1d).

In the early 1990s, another technology with incredible potential gained the attention of biomedical research: that is, *microarrays*, which are chips that, intuitively, allow for the capture of information about genes that can be used to identify groups with common behavior in order to infer, for instance, the level of involvement of each group in the same pathologies. Almost simultaneously with the announcement of the completion of the human genome, have appeared in the same or in equally prestigious scientific journals, studies on the automatic classification of tumors, although there has not been much media coverage about them. The fundamentally new proposal in those studies is to produce accurate tumor taxonomies via gene expression experiments with the use of microarrays. Those taxonomies, in turn, are the initial point of research efforts that, in the future, would allow for the focusing of treatment of the specific pathology affecting a given patient. It is not a coincidence that microarrays are also a fundamental tool for drug design and discovery.

From the 1990s to today, thanks to the reduction in cost, both large-scale sequencing technologies and microarrays are part of the investigation tools of research institutions, even small ones. Such a widespread use of those so-called high-throughput technologies has resulted in data production in quantities such as to cause serious management problems both for data warehousing and analysis. Therefore, as a consequence, there has been an exponential growth both of specialized databases for biological research and of computer science tools essential for the analysis of those data.

Computer science, mathematics and statistics are therefore fundamental, certainly for the data warehousing aspects, but even more for the analysis of those data in order to reach conclusions of relevance for biological research. Algorithmics has already given contributions that are at the base of tools now recognized as essential for the life sciences, such as the program BLAST, which is of daily use for sequence database searches. Here an effort is made to give the reader an idea of those contributions, limiting the scope to some of the areas of the biomedical sciences where research is particularly intense and where the computational techniques have not yet reached their full potential. It has already been established, in the previous chapters, that algorithmics, tightly connected to combinatorics, proposes automatic procedures to determine solutions to many computational problems that are based on deep theories and that try to shed light on what is information and how it is best represented and used. That will be exemplified in what follows by presenting some algorithms for bioinformatics that are recognized as reference points, either because they have been evaluated to be the best in a particular domain or because they have received particularly prestigious honors, such as the front page of outstanding scientific journals, e.g., *Nucleic Acids Research* and *BMC Bioinformatics*.

7.2 The Fundamental Machinery of Living Organisms

> A monkey is a machine that preserves genes up trees; a fish is a machine that preserves genes in water; there is even a small worm that preserves genes in German beer mats. DNA works in mysterious ways.[1]

DNA and proteins are polymers, composed of subunits known as nucleotides and amino acids, respectively. The genomic DNA of an organism, by means of the genes contained in it, is the information dictating the working of a complex biochemical machine whose aim is to produce proteins. Such DNA does not exist as a nude molecule, but rather as an extremely compact, three-dimensional, protein-DNA complex, known as chromatin. This latter is obtained via a process known as DNA supercoiling (Fig. 7.2), briefly described in box "DNA as Beads on a String". Intuitively, chromatin is the structure that is obtained once a long string has been wrapped around a series of beads in such a way as to take little space. The role of chromatin is not limited to such a compression process, as was initially thought, since it also has a deep influence on gene expression. Indeed, once packaged, only part of the genomic DNA is accessible and many messages are hidden. In order to transform those latter messages into proteins, they need to be made accessible. Such a goal is met via a dynamic behavior of the base components of chromatin. Here we limit ourselves to pointing out that research on such dynamic behavior is among the most important and fundamental in the life sciences because its understanding is seen as a substantial step forward for the cure of genetic diseases.

> **DNA as Beads on a String**
> Eukaryotic DNA can be seen as a very long thread. In order for this thread to be contained in the nucleus of a cell, it is necessary that it folds through a series of steps, at different levels of hierarchical organization, carried out through the use of particular proteins such as histones. The key steps of this process, known as supercoiling of DNA, are shown in Fig. 7.2. It is worth mentioning that only the first step of this process is known, while for the

(continued)

---

[1]Dawkins [25].

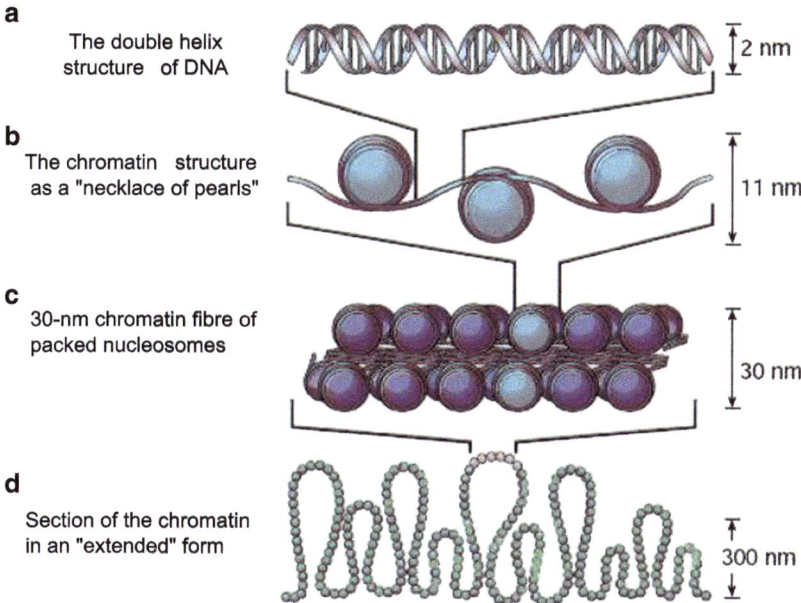

Fig. 7.2 The DNA supercoiling process

> **(continued)**
> others there are many working hypotheses about their structural conformation. The following explains in more detail the representations shown in the figure: (a) A DNA strand. The winding of DNA around spool-like structures called histones. Each spool with DNA wrapped around it is referred to as a nucleosome. (b) The resulting structure after this first step of packaging can be seen as a necklace of pearls, where the pearl is represented by the nucleosome. Note that the nucleosome is a fundamental unit in this process of compaction and bending, as DNA is wrapped at regular intervals, around histones to form nucleosomes. (c–d) The nucleosomes are then further packaged.

There is an estimate that each cell of a human being contains about 30,000 genes, that are present at birth and that remain stable throughout the entire life of an individual. Depending on various circumstances, including pathological ones or reactions to drugs, each gene is either activated or deactivated. In order for a gene to become active (technically, expressed), one needs to use "switchboards" referred to as promoters: DNA sequences that, on a genome, usually precede the DNA sequences corresponding to the genes. In each promoter, there are some "switches" that need to be "turned on" by some very special and important proteins,

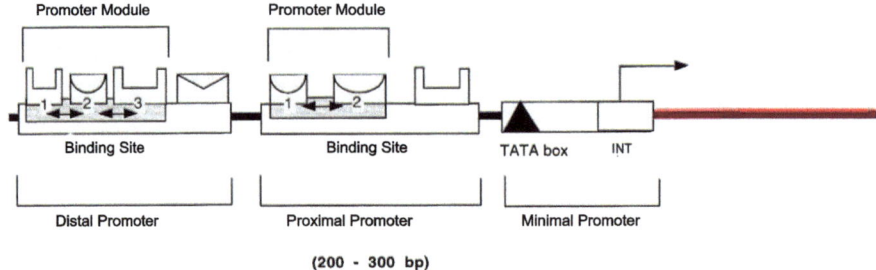

Fig. 7.3 The structure of a regulatory region

known as transcription factors. This entire process, known as transcription, is briefly illustrated next, starting with the description of the organization of a typical genomic region involved in the process, and schematized in Fig. 7.3. The region of interest can be divided into two smaller regions: "coding" (thick line) and "regulatory", where the first is to the right of the second. In the regulatory region, there are three promoters: the minimal promoter, very close to the start of the coding region, the proximal, which is further apart, and finally the distal, which can be even thousands of bases upstream of the beginning of transcription. That is not problematic since DNA is a dynamic three-dimensional structure and therefore sequences that are very far from each other may be close, or brought close to each other, in three-dimensional space. Some binding sites are indicated within each promoter.

When a binding site is "occupied" by a transcription factor, the effect is to recruit the RNA polymerase that begins the real transcription process. With reference to Fig. 7.4, the coding region is divided into introns and exons. The RNA polymerase transcribes the entire region forming the precursor RNA. Via a process known as splicing, some introns are removed to form the messenger RNA. This latter is then translated into an amino acid sequence corresponding to the desired protein.

The above transcription mechanism is common to all living species and it is therefore a fundamental one, whose malfunctioning may result in serious pathologies. Roger Kornberg, in 2006, received the Nobel Prize for Chemistry for his contributions to the understanding of the molecular basis of transcription. Those contributions could lead to the development of therapies, based on stem cells, for tumor and cardiovascular diseases.

The mechanisms and the processes that regulate gene expression and the quantity of proteins that result from that expression are extremely complex and much remains to be discovered and understood. One thing that is certainly clear is that malfunctioning of those expression mechanisms is at the origin of many pathologies. In order to shed light on the level of complexity of research in this area, we limit ourselves to mentioning that it has been discovered, only recently, that some small RNA sequences, known as microRNA, have a very important role in gene regulation, by inhibiting the production of some given proteins, de facto "silencing" or lessening the expression level of the corresponding genes. Recent studies, which

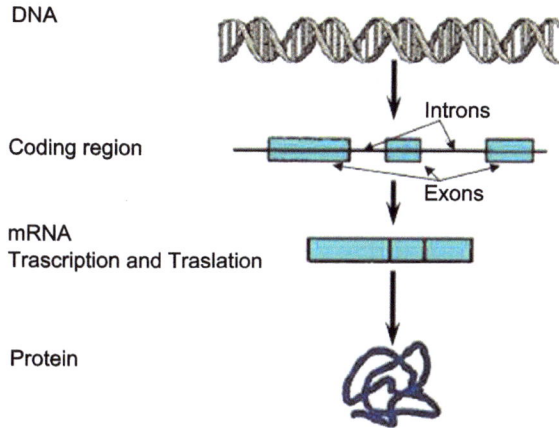

Fig. 7.4 The structure of a coding region and gene transcription

establish that microRNAs play a fundamental role in tumor development, also put forward a research plan for their use in therapy. It is also worthy of mention the way in which genetic information is transmitted to new cells. Indeed, that mechanism is what indissolubly connects the DNA molecule to inheritance and evolution. The double helix structure of DNA has already been discussed (see again the box "Textual Representation of DNA") and it is also well known that during cell division, only one helix of DNA is "passed on" to a new cell. In that new cell, DNA will appear again in its double helix form thanks to DNA polymerase which reproduces the missing helix from the one that is present. DNA recombination and copying are error-prone processes and therefore variations in a genomic sequence may be introduced. The simplest one is a substitution, consisting of one letter being replaced by another in a given genomic position. If that mutation is transferred to the offspring it enters the "big game of evolution". The human genome is made up of nearly three billion bases, and it is estimated that the difference between any two given genomes is on the order of about three million bases. Nature, which seems to appreciate combinatorics and make extensive use of it, leaves room for $4^{3,000,000}$ different human genomes. Moreover, although it would be nice to think of ourselves as being a "unique" species, 99 % of our genetic code is very similar to that of other mammals. In addition, many of our genes are similar to those of many other species, including fruit flies, worms and ... bacteria. In fact, winning biochemical mechanisms are preserved or, more precisely, the mechanisms that are preserved are the winning ones. For instance, histones are among the most conserved eukaryotic proteins and that is a clear indication of their involvement in fundamental biological processes. A guiding principle that one can abstract is that, even in biodiversity, genomic and proteomic similarity is a notable indication of biological relevance. Such a principle gives rise to one of the fundamental working hypotheses of computational biology:

> Similarity of genomic or proteomic sequences and structures, as measured by suitable mathematical functions, is a strong indication of biological relatedness, in evolutionary and/or functional terms.

Equipped now with that working hypothesis, apparently very fragile given the complexity of the "machine of life", we will now enter into some areas where algorithmic research has obtained some very valuable successes.

7.3 Algorithmic Paradigms: Methodological Contributions to the Development of Biology as an Information Science

> ...The discovery of DNA structure started us on this journey, the end of which will be the grand unification of the biological sciences in the emerging, information-based view of biology.[2]

A genomic sequence contains two types of digital information, suitably represented: (a) the genes encoding the molecular machinery of life, the proteins and the RNA, (b) the interaction and regulatory graphs that specify how these genes are expressed in time, space and intensity. Moreover, there is a "hierarchical flow of information" that goes from the gene to the environment: gene → protein → protein interactions → protein complexes → graphs of protein complexes in the cell → organs and tissue → single organisms → populations → ecosystem. The challenge is to decipher what information is contained within this digital code and in the hierarchical flow that originates from it. Since a few years after the discovery of DNA structure and the first sequencing experiments, algorithmics has played a key role in that decoding process. In order to point out the impact that such algorithmic studies have had on the life sciences, both in terms of tools and methodologies, it suffices to mention the following two examples. The BLAST program, the result of deep studies combining statistics and algorithmics, is a working tool that is now indispensable for the analysis of biological sequences. The sequencing of a genome by the shotgun sequencing techniques is now a consolidated reality, but it had a rather controversial beginning. One of the first studies to clearly indicate the feasibility of that type of sequencing on a genomic scale is based on algorithm theory. In what follows, we briefly present algorithmic paradigms, i.e., general approaches, that have made fundamental contributions in several areas at the forefront of research in the life sciences.

---

[2]Hood and Galas [61].

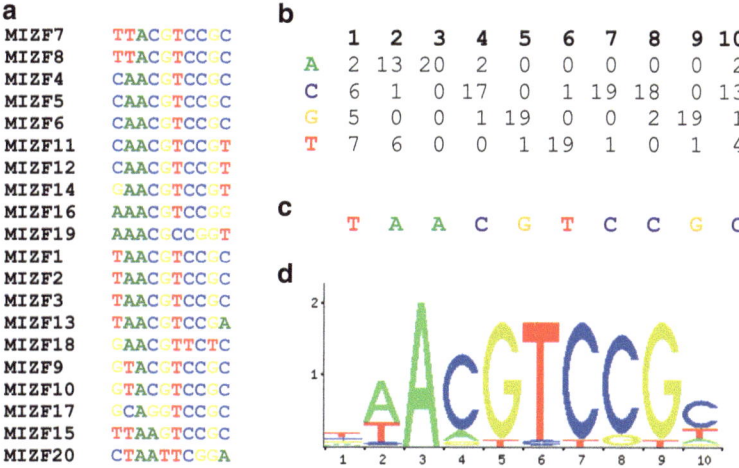

Fig. 7.5 Binding sites and motif representation

7.3.1 String Algorithmics: Identification of Transcription Factors Binding Sites

From the brief introduction given in Sect. 7.2, it is clear that one of the main areas of research in molecular biology is the discovery and understanding of the mechanisms that regulate gene expression, which have a strong implication also for medicine. To this end, an important line of research is the identification of regulatory regions, in particular binding sites of transcription factors. Obviously, carrying out such a discovery process purely via computational methods would be a great result for several reasons, e.g., the cost of experiments in silico compared to those in vitro. Unfortunately, this problem is difficult for many reasons, the main one being that the sequence that corresponds to a specific binding site is very short, 8–10 bases, and its retrieval could encompass the scrutiny of very long sequences, e.g., thousands of bases. It would be like looking for a needle in a haystack.

Transcription factors, in order to do their job, must bind to a particular DNA region. It is quite common that a transcription factor binds to a set of sequences instead of a single specific one. Those sequences are in different parts of the genome and usually share some characteristics that differentiate them from the other sequences in the genome. These common characteristics make it possible to describe these sites by a "motif", which can be defined in different ways. For the convenience of the reader, the notion of motif will be exemplified. Figure 7.5a shows the alignment of 20 sequences of binding sites of the transcription factor MIZF (zinc finger), i.e., a superimposition of the sequences summarized in a table. Figure 7.5b represents a Position Weight Matrix, i.e., a matrix of nucleotide frequencies in the positions of the alignment. Figure 7.5c shows the pure majority system: the motif is obtained as the consensus sequence from the matrix of frequencies electing the character with a relative majority in each column. Figure 7.5d exemplifies the pure proportional system: the motif is represented as a "logo" of the sequence obtained again from the matrix of frequencies: each character has a height, in the interval [0, 2], proportional to its frequency at that position. In general, given a specific transcription factor, the mining of its binding sites consists of finding all the shared sequence features in the site sequences. The main problem is the vague knowledge of the exact positions of interest in the genome. In particular, this knowledge is usually represented by a set of sequences, each sequence in the set corresponds to one or more binding sites.

The direct approach to the problem of extracting motifs from sets of sequences (each one several hundred bases long) offers only solutions based on enumeration and therefore expensive in terms of time. Fortunately, by taking advantage of statistical information about the relevance of each candidate motif, it is possible to reduce the search space, substantially improving the computational time of the algorithms.

A recent study has evaluated the best-known algorithms in the literature (11 as of 2005), on a benchmark dataset consisting of sequences with known binding sites. The performance of an algorithm is evaluated based on the percentage of binding sites correctly identified. Among the 11 algorithms examined, the best is Weeder, an algorithm that uses, in a very brilliant and original way, data structures and statistical counting techniques representative of many algorithms designed for the mining of textual information. The core of the algorithm is the suffix tree, a ubiquitous data structure used to represent textual information. In the area of data structures, the suffix tree is one of the most fundamental and useful ones: it has been developed from basic research and is largely used in several applications in bioinformatics. (The box "The Suffix Tree Data Structure" shows an example of a suffix tree and details some of its features.) A few years later, the algorithm MOST was developed to solve the same problem. While differing from Weeder, the core of MOST is still a data structure analogous to the suffix tree. The main idea of both algorithms is the identification of portions of a genomic sequence that are "over-represented" or "under-represented", i.e., portions of a sequence that are repeated more frequently or less frequently than expected. In fact, sequences that have an abnormal "statistical behavior" usually also have an important biological function. The computational complexity of the above algorithms depends, strongly, on the particular instance in input. For instance, for Weeder, the computational time can take from a few seconds to several hours.

The Suffix Tree Data Structure

A suffix tree is a data structure designed to represent a sequence of characters, highlighting the suffixes that comprise it. More in detail, a suffix tree for a given sequence S of n characters is a rooted tree (see Chap. 2) with n leaves.

(continued)

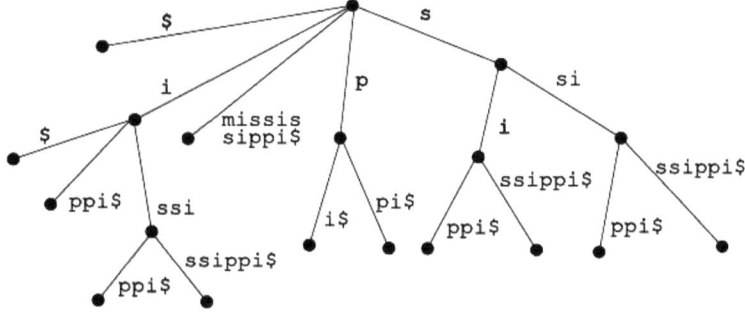

Fig. 7.6 The suffix tree for the sequence mississippi$

(continued)

Each internal node has at least two children and is labeled with a subsequence of S. The concatenation of the labels of the arcs of a path from the root to a leaf determines a specific suffix of the sequence S. To better illustrate this data structure, consider for example the sequence mississippi$. The $ has the formal role of preventing that a suffix is a prefix of another suffix. For example, ignoring the $ symbol in the sequence, the suffix i is the prefix of the suffix *ippi*, while the suffix $i$$ is not prefixed by any of the suffixes in mississippi$. This property allows for the association, one-to-one, between each suffix and its starting position in the sequence and it is essential for what follows. For example, the suffixes ssissippi$ and ssippi$ correspond to the positions 3 and 6 in the sequence. The suffix tree stores all the suffixes of a given sequence, such that: (a) suffixes with common prefixes share a path from the root to the leaves; (b) for each suffix, there is only one path from the root to a leaf associated with it, and vice versa. Property (b) is a direct consequence of the fact that, by construction, no suffix is a prefix of another in a given sequence. Thanks to the properties (a) and (b), the suffix tree stores all the subsequences of a sequence S and can be used to collect efficiently many statistics about S. Moreover, since suffixes that have prefixes in common share a path from the root of the suffixes tree, one has that identical subsequences will share a path starting from the root. In the suffix tree of Fig. 7.6, the root has five children because there are five different characters in the sequence that we are analyzing (including $). It is also easy to see that every letter appears in the sequence a number of times equal to the number of leaves in the subtree associated to it. Although it may seem surprising, there are algorithms linear in terms of computational complexity (i.e., complexity proportional to the length of the input sequence) able to construct a suffix tree.

7.3.2 Kolmogorov Algorithmic Complexity: Classification of Biological Sequences and Structures

The study of the evolution and classification of species has shifted from the consideration of morphological traits to the consideration of genomic ones. This change of approach has led, for example, to the discovery that the most common laboratory animal is not a rodent, even if it looks like it. Although there is a vast literature in the field and hundreds of algorithms have been designed for evolutionary studies, unfortunately the time performance of most of them does not scale well when it is required to classify very long sequences of bases or whole genomes, instead of sequences consisting of a few thousand bases. Below, we describe a solution to this problem which is particularly elegant, deep, and also effective. To this end, we need to introduce a classic notion of complexity that provides a very elegant measure of the "complexity" of an object, encoded by a text sequence x. This measure, $K(x)$, defined independently by Chaitin[3] and Kolmogorov,[4] and known as Kolmogorov complexity, is given by the length of the shortest program that produces x, without any input. Intuitively, the length of the shortest program required for the automatic construction of x provides a quantification of how complex x is. This insight and the corresponding definition can be extended in several ways. For example, $K(x|y)$ denotes the complexity of describing x (if y is known) and $K(x, y)$ denotes the complexity of describing both x and y. Kolmogorov complexity has several applications in a myriad of contexts, thanks to its links with statistics and the classic information theory founded by Shannon.[5] It is also a very elegant and simple way to quantify how two objects are "similar". In fact, if x is related to y, it is expected that $K(x|y)$ is smaller than $K(x)$. That is, if x and y are similar, the way to describe x starting from y is more concise than that of describing x, starting from nothing. Based on this observation, the theory of Universal Similarity Measures between two sequences has been developed and applied to the classification of sequences and biological structures, providing evidence of the validity of the approach. Further studies have clearly shown that the related biological tools based on this notion are extremely fast, scalable with the amount of data, and flexible. Therefore, they are very competitive with respect to the other previously known methods. For example, given a set of genomes it could be useful to group them in a hierarchical tree, depending on how similar they are based on their sequences. Since the similarity of genomes at the sequence

---

[3] Gregory John Chaitin is a well-known mathematician. When he came up with the idea and corresponding research on Algorithmic Information Theory he was only 18 years old and had just graduated from CUNY (City College, New York).

[4] Andrey Nikolaevich Kolmogorov was one of the greatest mathematicians of the twentieth century and perhaps that is the reason why this complexity measure carries his name.

[5] Claude Shannon is the founder of a mathematical theory of communication that has taken the name of Information Theory. This theory was born after World War II for a project regarding telecommunications networks, but it has had a very wide set of applications also in other fields.

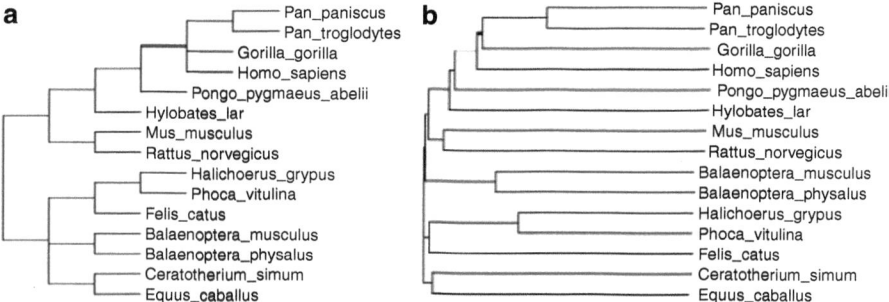

Fig. 7.7 The tree in (**a**) shows the taxonomy of the National Center for Biotechnology Information (NCBI), obtained based on biological considerations and found correct; the tree in (**b**) shows the classification of the same species obtained by using Kolmogorov complexity on mitochondrial DNA sequences

level is indicative of common evolutionary histories, it is expected that the tree produced should be a biologically acceptable phylogeny. To construct such a tree it is necessary to have algorithms that compute the similarity between genomes and then a rule that puts together the genomes, using the quantification of their similarities. This latter information is stored in a matrix, referred to as a similarity matrix, and its calculation is the most expensive step of the entire procedure. The algorithms that calculate similarities via Kolmogorov Complexity are considerably fast and accurate.

To illustrate this point, we consider 16 mitochondrial genomes, whose evolutionary classification is known and validated. The tree in Fig. 7.7a shows the taxonomy of the National Center for Biotechnology Information (NCBI), obtained based on biological considerations; the tree in Fig. 7.7b shows the classification of the same species obtained by applying the computational tools coming from Kolmogorov Complexity to the mitochondrial DNA sequences. The two trees are almost identical, not only "by eye", but also according to a formal mathematical similarity function between trees (the Robinson and Fould distance). In fact, the only difference is that the group of whales in the two trees do not have the same "close relatives". The construction of the second tree took a few seconds on a personal computer, while the first one was obtained using a semi-automatic procedure, involving also expert NCBI biologists who used knowledge available in the literature. Thus, not only is the automatic method fast, but it also provides a good starting point for the biologist to obtain a biologically valid classification.

7.3.3 Graph Algorithmics I: Microarrays and Gene Expression Analysis

In the introduction of this chapter, microarrays were mentioned as a technology that allows for the study of the level of expression of many genes, subject to the same

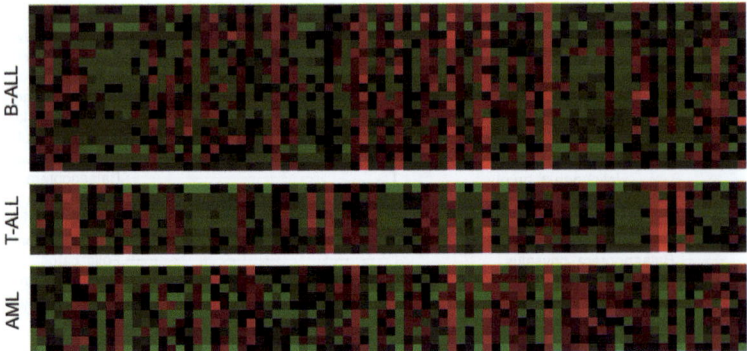

Fig. 7.8 The leukemia microarray for 38 patients (*rows*) and 100 genes (*columns*)

experimental conditions. This procedure gives very useful information in order to determine which genes have similar behaviors (expressed or non-expressed) under similar experimental conditions. The end result of an experiment using microarrays is a numerical matrix, referred to as an expression matrix. Generally, this matrix has a number of rows equal to the number of genes involved in the experiment and a number of columns equal to the number of experimental conditions. When microarrays are used for the molecular classification of tumors, the rows are associated with patients while the columns are associated with genes. The numerical value in the entry (i, j) quantifies the expression levels of gene j in a cell of patient i. A row gives a signature of how the genes behave in the same patient. It is expected that patients with similar diseases have "similar" rows in the expression matrix. The "clustering" of the patients depending on the similarity of the behavior of some genes could lead to the identification of more accurate methods for disease classification.

A brief example may help to clarify those ideas. Figure 7.8 shows, graphically, the gene expression matrix of a leukemia study. The rows are the histological samples of 38 patients and the columns are expression levels of 100 genes, carefully selected from over 16,000 that have been studied for that pathology. The patients are divided into three classes, depending on the type of leukemia affecting them: AML and ALL, this latter being further divided into two groups, the lines T-ALL and B-ALL. Given in graphical form, the different tones of the image correspond to different expression levels. The microarray shown in Fig. 7.8 is part of a study about the classification of tumors on a molecular basis. This analysis led to the construction of computer procedures that are able to accurately diagnose the type of leukemia without the intervention of experts. Therefore, it has produced an automatic diagnostic tool of great help to doctors, particularly those who have no significant experience with that particular disease. This study is the first establishing that it is possible to build diagnostically accurate and clinically relevant tools for the classification of tumors, using microarrays and experimental computational techniques.

From the computer science point of view, one of the main fundamental problems for the analysis of microarray data is the "clustering", i.e., the division of the rows of the expression matrix into "similar" groups, referred to as clusters. Although clustering is a widely studied problem, data from microarrays are very difficult to analyze in this context. This is due to the fact that the rows of the expression matrix are considered as vectors in a high-dimensional geometric space, making this problem very difficult. This situation has revived interest in the development of clustering algorithms, specific for gene expression data. Two of the most recent and distinguished clustering algorithms are Cast and Click. They have an important role in various research topics and are provided in all major analysis platforms for microarray data available in the literature. Following the literature, both algorithms consider the problem of clustering as one of partitioning[6] a given set into disjoint subsets. Formally, the set to be partitioned is a graph that has a vertex for each object to be classified. Then, for each pair of objects (i, j), there is an edge labeled with a certain weight, i.e., a real number which measures the "similarity" between objects i and j, as given by the corresponding rows of the expression matrix. The two algorithms use different techniques for partitioning the graph into subgraphs. Click tries to get the subgraphs formalizing clusters as a problem of network flow, where the edges with a low weight are removed. The theoretical version of Cast tries to build subgraphs as close as possible to complete subgraphs, or subgraphs that do not miss any of the possible edges. Starting from the purely theoretical Cast version, it has been possible to derive a powerful heuristic process for clustering gene expression data, which has the same name and is shown in the box "The Main Steps of Cast".

The Main Steps of Cast

Let s be a similarity function between two objects x and y, with values in $[0, 1]$—the greater the value of $s(x, y)$, the more similar the objects are; let $S(x, C) = \sum_{y \in C} s(x, y)$ be the total similarity between an element x and a set of elements C, where α is a discrimination parameter with values in $[0, 1]$. Cast identifies a cluster at a time via an iterative process. Assume that the following "status" holds in a generic iteration of the algorithm: there are elements in the list UC that have not been clustered yet and a partial cluster $Ctemp$ is under construction. $Ctemp$ is modified by two basic steps: ADD and REMOVE. They are performed in the order given and repeated until one can no longer add or remove items from $Ctemp$. At this point, $Ctemp$ is declared stable and labeled as a cluster itself. The procedure resumes with

(continued)

---

[6]The term partition refers to a decomposition of a set of "items" into disjoint subsets, whose union is equal to the entire set.

> **(continued)**
>
> *Ctemp* empty, in the case that there are still elements to be clustered (*UC* is not empty). In the following, the ADD and REMOVE steps are detailed.
>
> ADD: an element x is chosen from *UC* such that $S(x, Ctemp) \geq \alpha|Ctemp|$ is maximized. That is, the similarity average of x with elements in *Ctemp* must be at least α percent and maximal. If that condition is satisfied, x is included in *Ctemp*, and removed from *UC*.
>
> REMOVE: an element y is chosen from *Ctemp* such that $S(y, Ctemp) < \alpha|Ctemp|$ is minimized. This means that the average similarity of y with elements in *Ctemp* is below α percent and minimal. If that condition is satisfied, y is included in *UC* and removed from *Ctemp*.

Click and Cast are extremely fast and take a few seconds on microarrays with hundred of thousands of genes and conditions, usually providing the number of groups in which it is reasonably possible to divide the set of genes. Finally, we briefly come back to the leukemia data. Experimentally, it has been verified that, given an expression matrix and an appropriate choice of the input parameters, Cast is able to reconstruct the classification in Fig. 7.8, with only two exceptions. It is worth pointing out that such a dataset is "simple" to cluster while, in other circumstances, one must be prepared to have a far lower percentage of accuracy.

7.3.4 Graph Algorithmics II: From Single Components Towards System Biology

Several relevant studies indicate that the identification of interactions between different components, such as proteins, at the level both of single and different organisms, plays a fundamental role in biology. In the following, we briefly highlight that such interactions are of several types and all of them are very important for the identification of cellular machinery and evolution histories that characterize and differentiate species. For example, the human being and the chimpanzee are very similar, both in terms of genome sequences and gene expression levels. However, the interactions between genes (graphs of genes) are very different in the two species, in particular regarding the central nervous system. Like modern electronic devices, many components in humans and chimpanzees are similar, but the difference is in the "circuitry" which determines how these components interact. Discovering the similarities and differences of this circuitry, among various species, provides important information for system biology, where one of the main goals is the identification and the understanding of the fundamental properties of those interactions at the biomolecular "system" level. The significant amount of biological data produced in the last few years has uncovered several biomolecular interactions that can be encoded as graphs, both for the human being as well as for other model species. In the past 5 years, the number of interaction graphs available has increased by more than one order of magnitude. Moreover, technological advances will allow an exponential growth in the number of such graphs. Such a growth of the available interaction data is similar to that seen in the 1990s for genomic and proteomic sequences. The main difference is that the analysis of genomic and proteomic sequences had at its foundation over 40 years of research in algorithmic theory, while there is no such a background for interaction graphs.

In the literature, there are several interaction graphs of biological interest: protein–protein, gene regulation, co-expression and metabolic. Each of those graphs has its own features, but all have in common two computational problems whose solution, when mature, will provide research tools to the life sciences as important as the BLAST tool. The first problem is to identify, given a set of graphs, the common structure shared by them. That is, subgraphs that appear similar in all graphs. This problem is very similar to, but much more difficult than, the identification of patterns in sequences (see Sect. 7.3.1). In fact, it can be phrased again as a motif discovery problem, but this time in terms of graph structures. The other problem is: given a "query" graph, identify all graphs in a database in which there appear subgraphs similar to the query graph. Although there is a characterization of the mentioned two problems in terms of computational complexity, their study for the design of algorithms for biological graphs is in its infancy.

In what follows, we provide a short description of NetMatch, an algorithm that, given a "text" graph and a "query" graph, identifies in the text graph the subgraphs similar to the query graph according to a precise notion of similarity.

For example, consider the protein–protein interaction graph of yeast (*Saccharomyces cerevisiae*), annotated with gene ontology, shown in Fig. 7.9a. Assume one wants to find, in that graph, paths starting from proteins localized in the plasma membrane to proteins in the nucleus, passing through kinase proteins. This request is encoded by the query graph shown in Fig. 7.9b. Note that the nodes in the graph in Fig. 7.9b are connected by dashed edges. This encodes the fact that these edges can be replaced by paths during the search process. The graph in Fig. 7.9c represents an answer to the query. From the algorithmic point of view, the problem addressed by NetMatch (isomorphism between subgraphs) is computationally difficult, i.e., it is an NP-complete problem and is conjectured to have an exponential time complexity (see Chap. 3). Since the isomorphism between subgraphs is a fundamental problem in many contexts, different heuristic solutions have been studied. NetMatch generalizes to the case of "approximate isomorphisms" some of the known algorithms in the literature for the exact solution. In particular, in order to have a fast program on biomolecular graphs, several engineering speed-ups have been used. In fact, the program is able to perform in a few seconds approximate searches on complex interaction graphs.

Fig. 7.9 An example of protein interaction search in a biomolecular circuit. (**a**) A protein–protein interaction graph; (**b**) the "query" graph; and (**c**) result of the query

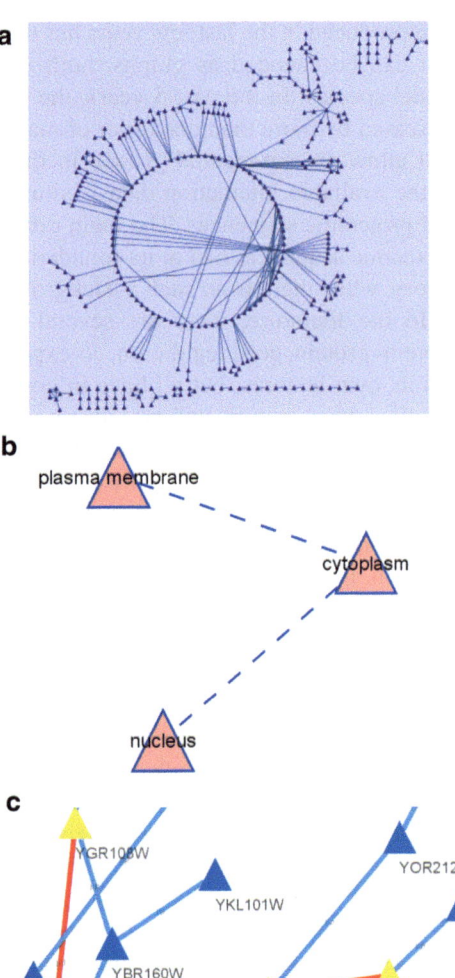

7.4 Future Challenges: The Fundamental Laws of Biology as an Information Science

The paradigm shift, just started, leading to the extension of the information sciences to biology raises important questions that will affect both the biological and the information sciences. It is quite clear that this new branch of the information sciences has a multidisciplinary nature, and biologists, physicists, chemists, computer scientists and mathematicians will play a key role. It is also clear that from this new process a new science and technology will arise, and we have seen only the beginning of it. Such a process requires a huge cultural change, even in the way of teaching all the involved disciplines to new students wishing to contribute to the development of this new branch of the information sciences. On this basis, the classic information sciences have to solve a first great challenge: the characterization of the complex biological information in terms of mathematics and computer science. That is, given the stochastic nature of biological processes we want to discriminate the "relevant biological signal" from the "noise introduced by stochastic processes" in biological data. A brief overview of past efforts could help us to understand what we need in the future. Turing and Shannon, between the 1930s and the 1940s, developed theories that reveal some of the basic laws for the transmission and processing of information, which have led to the development of the foundations that are the basis of the modern "information society" (see Chap. 5). The revolutionary contribution made by those two scientists was to show that something as impalpable as information could be defined, quantified and processed with mathematical tools. Over the years, those theories have become more refined and sophisticated. Moreover, they have led to the development of practical tools used in the transmission of signals and data processing. To all of this, algorithmics has provided vital contributions not only by introducing paradigms but also by proving that there are some intrinsic limitations on how efficiently a problem can be solved by a computer. However, despite the already great knowledge the information sciences have, it does not seem sufficient to provide adequate tools for the characterization and interpretation of "biological complexity". Indeed, the issue of *Science* dedicated to the human genome sequence mentions in its conclusions about ten notions of mathematical complexity known in the literature, but none of them seems to be suitable to characterize "real" biological complexity. The definition and use of this new notion of complexity seems to be a main priority for an information-based approach to biology. Algorithmics for the life sciences can only take advantage from such a foundational support in order to establish the complexity and the amount of information contained in a biological system. On the other hand, it can contribute to the development of this new notion by providing increasingly more accurate research tools to identify biologically meaningful events in the current information overflow characterizing the life sciences.

7.5 Bibliographic Notes

The impact that the discovery of the double-helical structure of DNA has had on science and culture is well presented in a *Nature* special issue that celebrates the 50 years from the discovery of DNA [84]. The essay by Lander [70], although a bit dated, presents the challenges for post-human-genome genomics that are still current and also proposes a global view of biology. Algorithmics has given outstanding contributions to genomic large-scale sequencing, including the human genome, that have not been presented here. Those aspects are presented in [48].

The importance of chromatin in controlling gene expression is well presented in a classic paper by Felsenfeld and Groudine [38], while the involvement of miRNA in cancer was addressed by Calin and Croce [14]. The paper by Hood and Galas [61] is part of the already-mentioned *Nature* special issue. The strong evidence that the guinea pig is not a rodent is presented in D'Erchia et al. [27]. In regard to introductory textbooks presenting algorithms for bioinformatics, those by Gusfield [53] and Jones and Pevzner [63] are worthy of mention. The BLAST sequence alignment algorithm is described in both books. The importance of the suffix tree in bioinformatics is presented in [54]. An elementary presentation of motifs in DNA sequences is given in [29]. Weeder and MOST are presented in [90] and [91], respectively. Two of the papers that develop the research line of classification through Kolmogorov complexity are [40, 73]. More generally, Kolmogorov complexity theory and its applications are well presented in [72]. The study about the leukemia data and the subsequent development of automatic classifiers for such a pathology is presented in [51], while the Click and Cast algorithms are presented in [99], together with many issues regarding microarray data analysis. Fast algorithms for internal validation measures as they apply to microarray data analysis are described in [49]. The state of the art regarding biomolecular graphs is presented in [100, 115]. Finally, NetMatch is described in [41].

Acknowledgements The author is deeply indebted to Luca Pinello and Filippo Utro for helpful discussions and comments about the content of this chapter. Many thanks also to Margaret Gagie for the usual, very competent proofreading and stylistic comments.

Chapter 8
The Shortest Walk to Watch TV

Fabrizio Rossi, Antonio Sassano, and Stefano Smriglio

Abstract The problem of finding the shortest path between two points underlies the concept of distance. In the common understanding, the physical distance between two points is always regarded as a non-negative quantity. However, from a mathematical point of view, the shortest-path problem can be defined and solved even when distances between points are negative. In this chapter we show that this model has an engineering application in the problem of synchronizing several electromagnetic signals received by a set of antennas. Solving this problem is fundamental in the design and implementation of digital television networks.

8.1 A Different Idea of Television

Our idea of television is intrinsically associated with content: news, reports, movies, shows, this is what we usually call "television". Occasionally, we happen to take notice of other aspects, related to the quality of the image and sound. For instance, if the reception is poor, we try to rotate the antenna or install some kind of amplifier to improve the "useful" signal. Going further, we might come across several technological issues, such as electromagnetic propagation, signal coding, transmitting and receiving electronic devices. But what we are really not used to considering as like television is everything concerning the optimal design of transmission networks along with the related algorithms. In fact, problems and skills arising in this context have been historically hidden by the redundancy of

---

F. Rossi (✉) · S. Smriglio
Dipartimento di Informatica, Università dell'Aquila, via Vetoio, 67010 Coppito (AQ), Italy
e-mail: fabrizio.rossi@univaq.it; stefano.smriglio@univaq.it

A. Sassano
Dipartimento di Ingegneria Informatica, Automatica e Gestionale, Sapienza Università di Roma, via Ariosto 25, 00185 Roma, Italy
e-mail: sassano@dis.uniroma1.it

the resource hosting video transmission: the *frequency spectrum*. For instance, the design and the implementation of the first Italian broadcasting network, carried out in the early 1950s, was straightforward. Having all the frequencies available, the engineers of the public broadcasting company (RAI, Radiotelevisione Italiana) accomplished the whole territory coverage simply by choosing suitable sites on hills and then setting up transmitters with the best frequencies in terms of service area. Nowadays the frequency spectrum is a public and very scarce resource. It also has a remarkable economic value, thanks to the increasing importance of wireless communications. To give an idea, the Spectrum Framework Review (Ofcom) evaluated that in the United Kingdom it was about 24 billion pounds in 2005. Other estimates show that it is about 2 % of GDP and employment in EU. No precise evaluations of the spectrum value in the Italian market have been carried out. However, an indication is provided by the 2 billion Euros revenue yielded to the Italian State by the auction of the 5-MHz bandwidth for UMTS services as well as the 136 million Euros from WiMAX frequencies. As a consequence, the optimal design of transmission networks (television, GSM, UMTS, WiMAX, etc.) has become a strategic goal for governments and regulation authorities. Specifically, a transmission network has to be designed so as to maximize the coverage while using the smallest possible amount of spectrum. In this way regulators are able to know, with good approximation, the *maximum number of operators* who can simultaneously use a specific portion of the spectrum without degrading the quality of service provided.

The process leading to the determination of the optimal network for a specific technology (digital television, digital radio or WiMAX) in a specific portion of the spectrum is referred to as *network planning*. For example, sophisticated planning techniques have been adopted to carry out the National Plan for Broadcasting Frequency Allocation issued by the Authority for Communications (www.agcom.it) in 1998, which set to 17 the maximum number of analogue television networks feasible in Italy.

This chapter shows how the design and implementation of algorithms is fundamental to providing high quality and widespread television service. We start by introducing the main elements of television networks along with basic notions and tools for their practical planning. Afterwards, we illustrate the decision problems concerning with the optimization of several transmitter parameters with a particular focus on *transmission delays*, which are a distinguishing feature of the new digital networks. We then present a mathematical model for the optimization of transmission delays, which is finally used to develop a planning algorithm. This is interesting for two main reasons. First, it is applicable to all transmission networks based on the *Orthogonal Frequency Division Multiplexing* (OFDM) modulation scheme, which is used by different types of wireless networks (e.g., 3G, WiFi, WiMAX, DAB), and, above all, by the digital video broadcasting (DVB) currently being implemented in Italy and Europe. Second, the algorithm has a "soft" impact on the network, that is, it does not require a relocation or modification of the transmission equipment but only a reprogramming of the transmitters. Nevertheless, it may greatly extend the network coverage. Furthermore, the mathematical model and algorithm have the advantage of being explainable without resorting to an advanced mathematical formalism, while maintaining a high level of generality. In fact, the modeling process also applies when other transmitter parameters, such as transmission frequency or emission power, have to be optimized. However, in these cases, the treatment is mathematically more complex with respect to the case of transmission delays, even if it does not present relevant conceptual differences.

8.2 Designing a Broadcasting Network

In this section we describe notions and concepts such as system elements, computer representation and performance evaluation that play a central role in designing broadcasting networks.

8.2.1 The Physical Elements of the Network

A terrestrial television network consists of a set of transmitters which broadcast television programs simultaneously in a certain geographical area. We refer to a *transmitter* as the collection of all devices required to radiate the signal over an area ranging from a district of a few square kilometers to entire regions. The antennas are the most visible parts of transmitters, being installed on towers that can reach considerable heights (more than 100 m). The antennas are fed by a set of electronic circuits such as oscillators, modulators and amplifiers. Due to the complexity of the infrastructure and to environmental protection standards that regulate the maximum level of electromagnetic emissions, the transmitters cannot be installed just anywhere, but only at sites previously identified by the competent authorities. Therefore, transmitters' geographical coordinates are fixed and known a priori, whereas signal radiation depends on the transmitter configuration. The radiation diagram is determined by several parameters, including the height, shape and antenna orientation, the transmission frequency, the emission power in the various directions, the signal polarization and, distinctively for the new digital networks, the transmission delay.

Users receive the signal emitted from the transmitters using suitable receivers. A receiver consists of various electronic devices (tuner, demodulator, the television screen) connected to a receiving antenna. It is a common experience that the shape and placement (orientation) of the receiving antenna significantly affect reception quality. Thus, unlike the transmitters, which are localized in a few carefully selected sites with well-defined characteristics, the receivers (or users) are non-uniformly spread over the territory and may have different characteristics (a roof antenna considerably differs from a portable TV antenna). A receiver is said to be *covered* by the service if programs appear clearly on the TV screen.

> Designing a broadcasting network consists of choosing a configuration of transmitters with the goal of maximizing the number of receivers covered by the service.

Intuition suggests that all receivers can be reached with adequate signal strength (and, thus, covered) simply by increasing the emission powers of all transmitters. Unfortunately, this is often not possible due to the *interference*, that is, the physical phenomenon of attenuation of the electromagnetic wave intensity occurring when signals from different transmitters overlap. Thus, designing a large-scale broadcasting network requires an appropriate model representing the system and the interference among transmitting antennae.

8.2.2 Computer Representation

A computer representation of a broadcasting network consists of three distinct models:

1. A digital model of the territory;
2. An electromagnetic propagation model;
3. A receiver model (that is, a model for coverage assessment).

To represent the geographical area and the receivers distributed over it, a geographical database known as the *Digital Elevation Model*, (DEM, Fig. 8.1) is adopted. A DEM consists of a grid of squared cells overlapped on the territory. For each cell, the altitude above sea level, the number of inhabitants (potential receivers) and the characteristics of a generic receiver (type of antenna and orientation) are stored. The reason why the territory is decomposed into cells is that, if a single cell is small enough, it can be assumed that all the receivers placed inside it receive the signal in the same way, regardless of their actual position. Therefore, each cell behaves as a single reference receiver and is called a *testpoint* (TP, Fig. 8.1).

It is well known that in wireless networks the signal emitted by a transmitter arrives at the receivers with some fading. This happens due to the distance between transmitter and receiver (propagation fading), to the presence of obstacles (diffraction fading) and also to different types of surfaces encountered during the path (reflection and refraction fading). The mathematical rules that define the calculation of the power received in a TP as a function of the emitted power and the signal path are referred to as propagation models. The International Telecommunication Union (ITU) is an international institute that defines the propagation models recommended for radio and TV networks. The output of a propagation model is a power value for each pair (transmitter, TP). Regardless of the specific model, let's see how many values have to be computed in a network with national extension that uses 2,000

8 The Shortest Walk to Watch TV

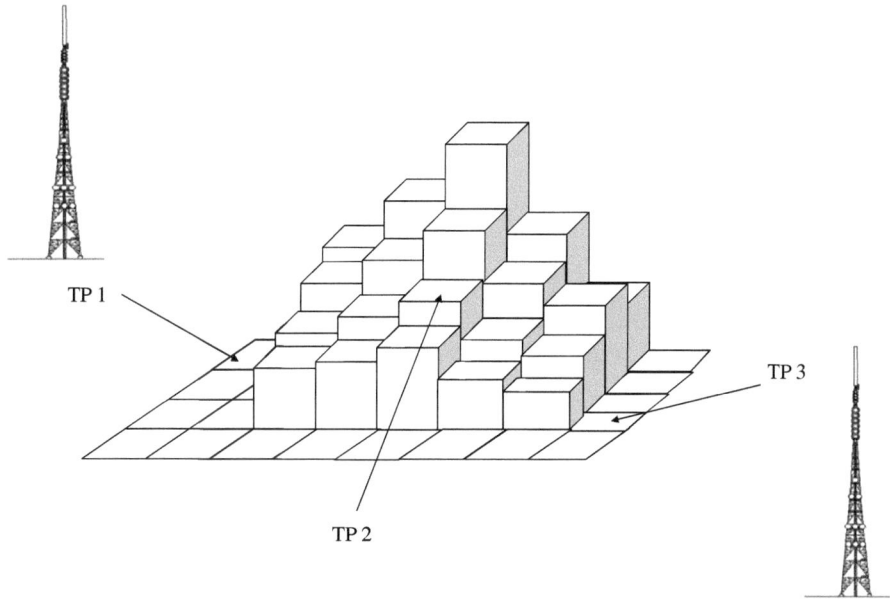

Fig. 8.1 Digital Elevation Model (DEM)

transmitters. Since the representation of Italy on a grid with cells of 250×250 m needs about 55,000,000 TPs, the propagation model should in theory calculate $55,000,000 \times 2,000 = 11$ billion power values! In practice, transmitters do not reach all TPs, but only some of them (think of a small transmitter in a valley surrounded by mountains). Assuming that 50 signals arrive on average in each TP, the evaluation of the received power for all TPs requires the calculation of about $55,000,000 \times 50 = 2.75$ billion power values. However, despite its complexity, this calculation is not yet sufficient to determine whether the television program is received correctly. To complete the coverage assessment we need a further model representing the receiver behavior and the way it handles the received signals.

8.2.3 Model for the Digital Coverage Assessment

The digital television signal, as with every numeric transmission, carries an encoding of the information associated with the programs: in practice, sound and video are encoded before being released in a continuous stream of symbols. A symbol transmitted by transmitter i at time t_i is received on TP j at time

$$\tau_{ij} = t_i + \Delta_{ij} \tag{8.1}$$

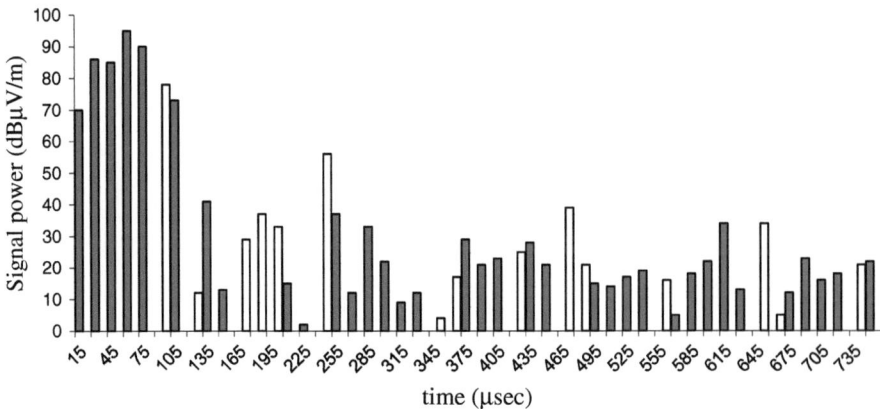

Fig. 8.2 Received signals at a sample TP

where Δ_{ij} equals the ratio between the distance from i to j (in km) and the speed of light (around 300,000 km/s).

The graph of Fig. 8.2 refers to a sample TP: it contains all the received signals corresponding to the same symbol transmitted simultaneously at time $t = 0$ from all the transmitters of the network. The signal arrival times (in microseconds, μs) are reported on the abscissa, while the ordinate measures the signal strength (in decibels microvolts/meter, dBμV/m), as computed by the propagation model.

The different shading of the signals represents the fact that they are transmitted at different frequencies (signals of the same color are transmitted at the same frequency). The receiver we are considering has the following properties:

1. It is able to tune only on one frequency at a time;
2. Signals received at a frequency different from the tuning frequency have a negligible effect on the quality of service.

In the example of Fig. 8.2, this gives rise to two different cases: in the first case the receiver will consider only dark signals, and in the second case will consider only the white ones. For each case (i.e., frequency), to determine if a TP is covered by the service at that frequency, one has to classify the received signals into *useful* and *interfering*. The first will contribute to the correct reception of images and sounds; the latter will tend to degrade their quality. Once the two families of signals have been identified, an aggregate power contribution is estimated for each of them, so as to obtain two quantities representing, respectively, the *total useful signal* and the *total interfering signal*. This estimate is carried out by a composition algorithm certified by the European Broadcasting Union, EBU. The ratio between total useful and interfering (augmented by the *thermal noise* of the system) signals is referred to as the *signal/noise* ratio. This quantity allows us to formulate the following coverage condition:

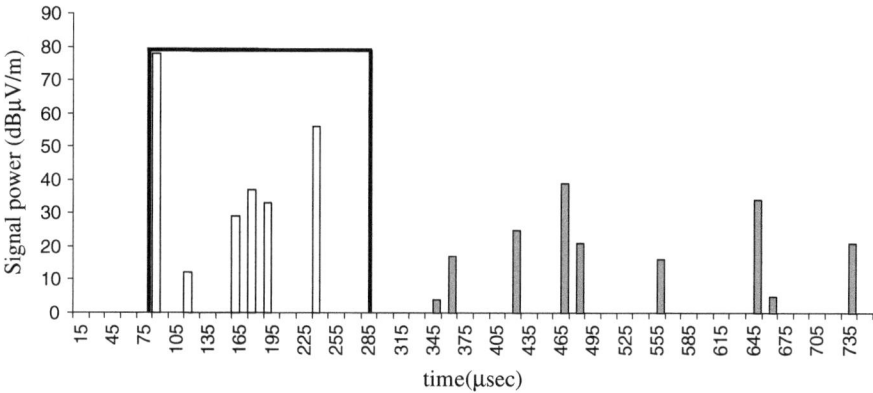

Fig. 8.3 Detection window

> A TP is regarded as being covered by the service if the *Signal/Noise* ratio exceeds a fixed value, which depends on the transmission technology.

The *classification scheme* of the signals used in digital terrestrial television refers to the standard DVB-T, Digital Video Broadcasting-Terrestrial. The DVB-T adopts a modulation scheme (the OFDM scheme) that allows the receiver to combine constructively isofrequency signals carrying the same symbol if they arrive not too far apart in time.

Specifically, any two such signals can be constructively combined if the delay of time distance between them is less than a given value T_G ($T_G = 224\,\mu s$ for current DVB-T receivers). The time interval in which the signals can be combined in a constructive way is called the *detection window*. In Fig. 8.3, a possible position of the detection window is represented on our sample TP, relative to the frequency of white color. All signals that fall within the window are combined in the total useful signal. The remaining signals with the same frequency, but now represented with dark color, contribute to the total interference signal.

In principle, the receiver can position the detection window at any point of the time axis. However, it can be proved that the significant positions are only those corresponding to the arrival times of the signals. Therefore, to check the coverage of a TP receiving k signals, only k positions of the detection window must be evaluated, each time discarding the signals at a different frequency from that on which the window is positioned. If the covering condition is satisfied in at least one of these positions, then the TP is covered by the service. Note that this introduces an additional multiplicative factor in the complexity of the previous paragraph: although to calculate the received power on the Italian territory we need 27.5 billion operations, to assess the coverage the number of operations grows to about $27.5 \times 50 = 1.3$ trillion!

8.2.4 Network Design

The general planning problem is trivially solvable if a sufficiently large number of frequencies is available: each time the activation of a transmitter at a given frequency creates interference one can assign a new (not yet used) frequency to such a transmitter eliminating all possible interferences.

Unfortunately, the spectrum of available frequencies is limited (TV networks in Italy can use at most 56 frequencies) and then its usage has to be optimized. Thus, to have the largest possible number of networks in a given area, each network must use the minimum number of frequencies. At the limit, by using just one frequency for all transmitters of the network, one may have as many networks as the number of available frequencies. This is generally not possible with traditional analogue networks. On the contrary, the reduction mechanism of the interference implemented in OFDM-based networks (described in Sect. 8.2) allows us to design single-frequency networks (*Single-Frequency Network, SFN*) with great extent, as shown for the first time in the *White Book on Terrestrial Digital Television*, produced by the Italian Authority for Communications in the year 2000. In fact, the features of the OFDM scheme can be exploited to enhance the total useful signal in TPs which are affected by interference. To clarify this fact, let us go through a network design by using only one frequency for all transmitters. It might be the case (and most likely will be) that on several TPs, even combining useful signals, the *Signal/Noise* ratio falls below the required threshold. Which "levers" does the designer have to improve the quality of service on such TPs?

One available option, which can lead to significant coverage enlargement, is changing the emission power of the transmitters. However, this can only be done by taking into account a variety of constraints: technology, mutual interference with other networks, coordination with neighboring countries, legislation for electromagnetic emissions, etc. Illustrating the resulting optimization problems and their solution algorithms would require a complex mathematical treatment, hardly accessible to non-experts. A second lever consists in introducing an artificial *transmission delay* to the symbol transmitted by some transmitters of the network. This operation is technically simple and has no effect on other networks. The next section describes the role of the transmission delay and the resulting optimization problem.

8.3 The Role of Transmission Delays

Let us consider a single-frequency network with transmitter set T and TP set Z. For each pair (transmitter i, TP j) we are given the following data:

1. The power emitted by i and received in j (computed by the propagation model);
2. The time Δ_{ij} required for a symbol transmitted by i to reach j, which equals the distance between i and j divided by the speed of light (0.3 km/µs).

8 The Shortest Walk to Watch TV

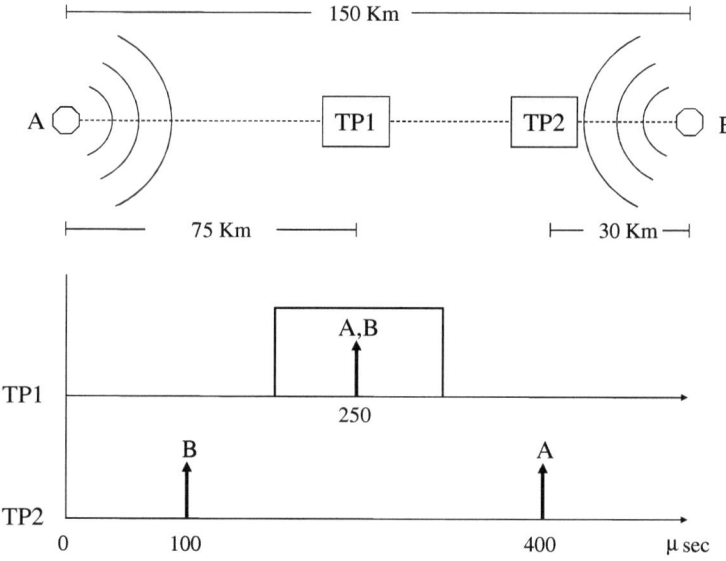

Fig. 8.4 Simultaneous transmission $t_A = t_B = 0$

Thanks to Eq. (8.1), we can compute the arrival times τ_{ij} as functions of transmission times t_i. For instance, in Fig. 8.4, we have two transmitters, A and B, 150 km away from each other, transmitting the same symbol in the same instant, that is, $t_A = t_B = 0$. The arrival times (expressed in μs) of the symbol transmitted from A are: $\tau_{A1} = 75/0.3 = 250$ in TP1, $\tau_{A2} = 120/0.3 = 400$ in TP2; while those from transmitter B are: $\tau_{B1} = 75/0.3 = 250$ in TP1, $\tau_{B2} = 30/0.3 = 100$ in TP2.

In the remainder of the treatment we make the following assumption:

> The received power values are such that any TP is covered by the service if and only if all signals carrying the same symbol fall inside the detection window.

In Fig. 8.4 this holds for TP1, but not for TP2. In fact, the received symbols on TP2 are separated by an interval of amplitude $|\tau_{A2} - \tau_{B2}| = 300$ μs, greater than the width of the detection window ($T_G = 224$ μs in the adopted standard). This means that there is no position of the detection window which allows TP2 to capture both symbols and, when $t_A = t_B = 0$, TP2 is not covered by the service.

One possibility for extending the coverage also to TP2 is to introduce a transmission delay in the transmitter A or B so as to bring the two symbols closer

Fig. 8.5 $t_A = 0$, $t_B = 100\,\mu s$: both TPs are covered

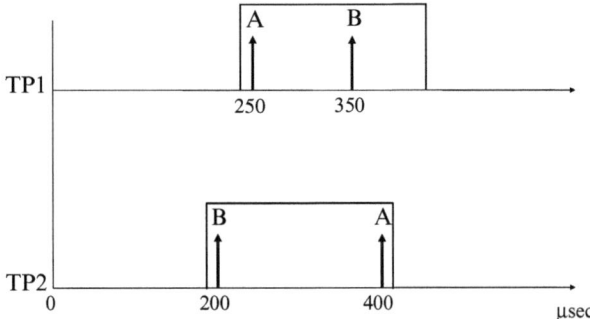

Fig. 8.6 A more complex network

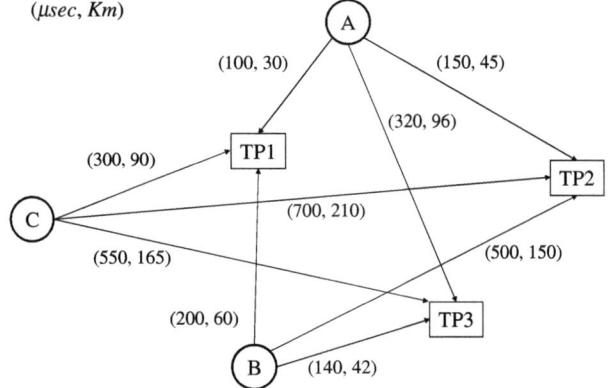

to each other and let both of them fall inside the detection window. Thus, we apply a delay $t_B = 100\,\mu s$ to transmitter B. The new configuration is shown in Fig. 8.5.

Since the delay is applied "to the source", that is, on the transmitter, the arrival instant of the symbol transmitted from B is delayed by $100\,\mu s$ on both TPs. Then, the received signals on the TP1 move away from each other. Nevertheless, the amplitude of the time interval separating the two symbols is less than the width of the detection window on both TPs. Therefore, there is a positioning of the windows that guarantees coverage of both TPs. It should be noted that, in principle, the same effect would be obtained by physically moving the transmitter B, specifically, by translating it to the right by 30 km. In conclusion, the introduction of a single transmission delay extends the coverage of the network. With increasing network complexity, the calculation of the delays that maximize the number of TPs covered cannot be done according to simple rules, but requires the use of more sophisticated algorithmic tools. To understand this, consider the network shown in Fig. 8.6, composed of three transmitters A, B, C and three TPs 1, 2, 3.

For each pair (transmitter, TP), the figures in brackets indicate, respectively, the propagation time (μs) and the distance (km). In Fig. 8.7 the arrival instants of the signals on each TP are represented in the case $t_A = t_B = t_C = 0$. Recall that,

8 The Shortest Walk to Watch TV

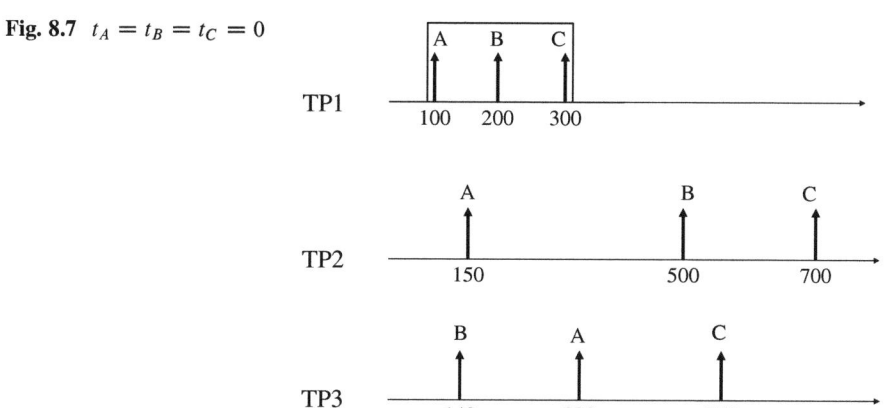

Fig. 8.7 $t_A = t_B = t_C = 0$

by assumption, a TP is covered if and only if all of the three signals fall within the detection window, that is, if and only if, for each pair of received signals, the time interval which separates them does not exceed the amplitude $T_G = 224\,\mu\text{s}$ of the detection window. This condition is not satisfied on TP2 and TP3. In fact, the time interval between the received signals A and B on TP2 is equal to $|\tau_{A2} - \tau_{B2}| = 350 > T_G = 224\,\mu\text{s}$; similarly, the time interval between the received signals B and C on the TP3 is equal to $|\tau_{B3} - \tau_{C3}| = 410 > T_G$. Thus, in the case $t_A = t_B = t_C = 0$, the only TP covered is TP1, thanks to the detection window represented in Fig. 8.7.

As a first attempt, we try to replicate the reasoning of the previous example, that is, we choose a transmitter and look for a transmission delay that induces an increase of coverage. We begin with transmitter A. By focusing on TP1, we observe that the maximum delay that can be assigned to A is imposed by the signal B. In fact, to force both of them to fall in the detection window, the arrival time of A cannot exceed $424\,\mu\text{s}$; therefore the maximum delay can be at most $324\,\mu\text{s}$. Similarly, TP2 is covered if the delay imposed on the transmitter A is at least equal to $326\,\mu\text{s}$, so as to move signal A close enough to C. Finally, we note that the TP3 is not covered because of the distance between the received signals B and C, and therefore a delay on the transmitter A has no influence on its coverage. Similar considerations hold for transmitters B and C. In detail, for each transmitter {A,B,C} and for each TP j, Table 8.1 reports the range of values that the delay of i can assume so that j is covered when the remaining transmitters transmit at time $t = 0$. It is easy to see that the introduction of a delay in transmission of a single transmitter does not allow us to obtain an increase of coverage. For example, the conditions for transmitter A on TP1 and TP2 are not simultaneously satisfiable. A similar situation occurs for the transmitter B on TP1 and TP3.

Nevertheless, the coverage of the network can still be increased by applying simultaneously a transmission delay to different transmitters. Specifically, say $t_A = 400\,\mu\text{s}$ and $t_B = 200\,\mu\text{s}$. The new configuration, represented in Fig. 8.8, shows how

Table 8.1 TPs coverage conditions with only one delayed symbol

TP	Condition on A	Condition on B	Condition on C
1	$0 \leq t_A \leq 324$	$0 \leq t_B \leq 124$	$0 \leq t_C \leq 24$
2	$326 \leq t_A \leq 574$	Never covered	Never covered
3	Never covered	$186 \leq t_B \leq 404$	Never covered

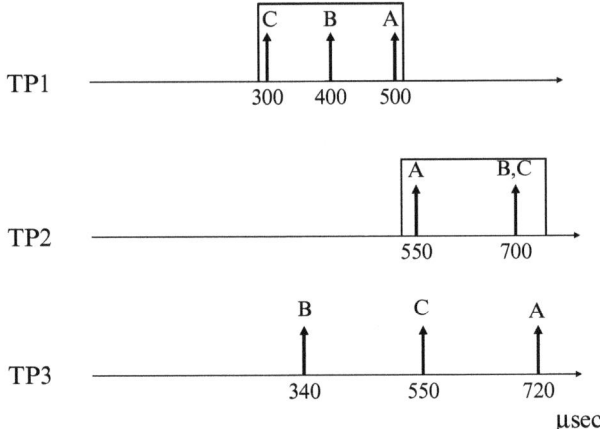

Fig. 8.8 $t_A = 400, t_B = 200, t_C = 0$

all three signals now fall within the composition window of TP1 and TP2, ensuring their coverage.

At this point, one may wonder whether, like the previous example, it is possible to determine an assignment of delays so as to cover all TPs. A close analysis of the diagram in Fig. 8.7 allows us to exclude this possibility. Let us focus on the signals A and B and TPs 2 and 3. For these signals to fall into the same window on TP2, signal A should be delayed by at least 126 µs. As for TP3, the two signals can fall in the same window only if A is delayed by at most 44 µs. The two conditions are not simultaneously satisfiable: no assignment of delays exists such that TP2 and TP3 are simultaneously covered.

In summary, the two examples show a fundamental difference. In the first example there is an assignment of the transmission delays so as to cover all test points in the given set Z. On the contrary, in the second example, only a subset of TPs in Z can be covered. In this case, our interest is to determine the maximum number of TPs that can be covered simultaneously by proper assignment of delays, that is, the largest coverage area of the network. This question leads to the definition of an optimization problem associated with the assignment of transmission delays.

Problem 1. Given a single-frequency network (T, Z) with fixed emission powers, determine an assignment of transmission delays to transmitters such that the number of covered TPs in Z is maximized.

In the remainder of the chapter, we develop a solution algorithm for this problem, which is derived by representing it on a suitable graph.

8.4 An Algorithm for Optimizing Transmission Delays

We have shown that there is no assignment of delays which covers simultaneously all the TPs of the network of Fig. 8.6. To prove this, we have identified a subset of TPs, precisely the set $\{2, 3\}$, which cannot be simultaneously covered. A set of this type is a *certificate* of non-existence of a solution that covers all TPs. We introduce, therefore, the following definitions:

Definition 1. A TP set Z is said to be *consistent* if there exists an assignment of delays such that all TPs in Z are covered by the service. If such an assignment does not exist, then Z is said to be *inconsistent*.

Definition 2. An inconsistent TP set Z is said to be *minimal* if, removing from Z any of its TPs, the resulting set S is consistent.

In the network of Fig. 8.6 the TP set $\{2, 3\}$ is a *Minimal Inconsistent Set* (MIS) and is also the only MIS contained in Z. If we remove from Z any one of the two TPs belonging to the MIS, for example, TP3, we obtain a new set $S = Z \setminus \{3\} = \{1, 2\}$ which is consistent (for example, with the assignment of delays $t_A = 400, t_B = 200\,\mu s$). In other words, by removing a TP from the unique MIS, we obtained a consistent TP set. The generalization to the case where Z contains more than one MIS is immediate: one TP must be removed from each MIS to obtain a consistent set. The algorithm, which proceeds by successive deletions of TPs (contained in some MIS), can be summarized as follows (Table 8.2):

Step 2 and Step 3.b need to be further specified: in particular, we must formally define a procedure to identify a MIS W in a TP set (Step 2) and a procedure for calculating the delays to be assigned to the transmitters if there is no MIS in S (Step 3.b). In the next section we show that these two tasks correspond to the solution of a single optimization problem.

Table 8.2 clean inconsistency algorithm

Input	a TP set Z
Output	a consistent TP set S
Step 1	Initialize $S = Z$
Step 2	look for a MIS W of TPs in S
Step 3.a	**if** W exists, **then** remove from S any TP contained in W; **goto** Step 2
Step 3.b	**otherwise** compute an assignment of delays such that all TPs in S are covered - **stop**

8.4.1 From Inconsistent TP Sets to Inconsistent Systems of Inequalities

Consider a TP set Z and recall that, by assumption, a TP j is covered if and only if, for each pair i, k of received signals, the time interval that separates them does not exceed the width of the detection window $T_G = 224\,\mu\text{s}$. This condition is expressed mathematically by requiring that the absolute value of the difference between the arrival instants of the two signals is not greater than T_G:

$$|\tau_{ij} - \tau_{kj}| \leq T_G$$

this is equivalent to a pair of linear inequalities:

$$\tau_{ij} - \tau_{kj} \leq T_G$$
$$\tau_{kj} - \tau_{ij} \leq T_G$$

let us now recall that the arrival instants depend on the transmission delays according to the expression $\tau_{ij} = t_i + \Delta_{ij}$, introduced in Sect. 8.2.3. We obtain:

$$(t_i + \Delta_{ij}) - (t_k + \Delta_{kj}) = t_i - t_k + \Delta_{ij} - \Delta_{kj} \leq T_G$$
$$(t_k + \Delta_{kj}) - (t_i + \Delta_{ij}) = t_k - t_i + \Delta_{kj} - \Delta_{ij} \leq T_G$$

Finally, moving to the right-hand side all constants, we have that a TP is covered *if and only if*, for each pair i, k of received signals, the following two inequalities are satisfied:

$$t_i - t_k \leq T_G + \Delta_{kj} - \Delta_{ij} = d_{ik}^j$$
$$t_k - t_i \leq T_G + \Delta_{ij} - \Delta_{kj} = d_{ki}^j$$

8 The Shortest Walk to Watch TV

Table 8.3 System of difference constraints defining the coverage conditions

TP1	$t_A - t_B \leq 224 + 200 - 100 = 324$	
	$t_B - t_A \leq 224 + 100 - 200 = 124$	
	$t_A - t_C \leq 224 + 300 - 100 = 424$	
	$t_C - t_A \leq 224 + 100 - 300 = 24$	
	$t_B - t_C \leq 224 + 300 - 200 = 324$	
	$t_C - t_B \leq 224 + 200 - 300 = 124$	
TP2	$t_A - t_B \leq 224 + 500 - 150 = 574$	
	$\mathbf{t_B - t_A \leq 224 + 150 - 500 = -126}$	
	$t_A - t_C \leq 224 + 700 - 150 = 774$	
	$t_C - t_A \leq 224 + 150 - 700 = -326$	
	$t_B - t_C \leq 224 + 700 - 500 = 424$	
	$t_C - t_B \leq 224 + 500 - 700 = 24$	
TP1	$\mathbf{t_A - t_B \leq 224 + 140 - 320 = 44}$	
	$t_B - t_A \leq 224 + 320 - 2140 = 404$	
	$t_A - t_C \leq 224 + 550 - 320 = 454$	
	$t_C - t_A \leq 224 + 320 - 550 = -6$	
	$t_B - t_C \leq 224 + 550 - 140 = 634$	
	$t_C - t_B \leq 224 + 140 - 550 = -186$	

The resulting system of inequalities is known as a *system of difference constraints*. This is exhaustively reported in Table 8.3 for the example of Fig. 8.6.

Now, we can state a relationship between the consistency of any TP set Z and the existence of solutions to the corresponding system of difference constraints. Specifically, the following property holds:

> *Property 1.* If the system of difference constraints associated with a given TP set Z admits solution, then the set Z will be consistent and its solution values will be precisely the transmission delays to be assigned; otherwise, Z will contain a MIS.

At this point, we need a solution algorithm which has the two possible outputs:

- The solution values, if at least one solution exists;
- A MIS, if the system does not admit solutions.

Let's go back to the network in Fig. 8.6. We know that its TP set Z is inconsistent, as it contains the MIS composed of TP2 and TP3, and we wish to detect such a MIS by an algorithm. If we look at the system of Table 8.3, we notice that it contains a subset (also minimal) of inequalities that can not be satisfied simultaneously. Precisely, the inequalities $t_B - t_A \leq -126$ and $t_A - t_B \leq 44$, associated, respectively, with TP2 and TP3 (marked in bold in Table 8.3), cannot be simultaneously satisfied by some nonnegative value of t_A and t_B. This is indeed a general rule:

> To detect a MIS in a TP set it is sufficient to identify an inconsistent subset of inequalities in the associated system of difference constraints. Such a subset of inequalities is itself said to be *inconsistent*.

In order to accomplish this task, we need again to resort to a different representation of the problem. Interestingly (and a bit surprisingly), we are going to represent the system of difference constraints through a directed graph $G = (N, A)$ with weights on the arcs.

8.4.2 The Difference Constraints Graph

The graph G is constructed as follows: the node set contains one node for each transmitter plus an additional special node s; each inequality $t_k - t_i \leq d_{ik}^j$ gives rise to an arc, directed from i to k, with weight d_{ik}^j. However, if a pair i, k appears in more than one inequality, then only one arc is actually included in A, namely, the one with minimum weight. The special node s is connected to all other nodes by zero-weight arcs. The graph corresponding to the system of Table 8.3 is drawn in Fig. 8.9.

If we now consider the arcs associated with inequalities $t_B - t_A \leq -126$ and $t_A - t_B \leq 44$ we observe that they form a directed cycle (A → B → A) with length (intended as the sum of the weights of the arcs), equal to -82. This is a critical structure for our purposes:

> *Property 2.* In general, it can be shown that inconsistent subsystems of the system of difference constraints correspond to negative-length cycles in G and vice versa.

Therefore, determining if a system of difference constraints contains an inconsistent subset of inequalities is equivalent to checking if G contains negative (length) cycles. Furthermore, if G does not contain negative cycles, then it can be used to calculate the delays to be assigned, that is, a solution to the system of difference constraints, as discussed in the next paragraph.

Fig. 8.9 Difference constraints graph for the system of Table 8.3

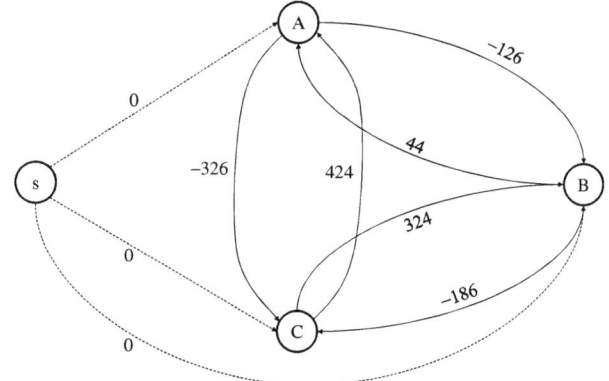

8.4.3 Shortest Walks in G and Transmission Delays

We define a *directed walk* in G as a sequence of nodes and arcs $\{n_1, a_{12}, n_2, \ldots, a_{k-1,k}, n_k\}$, where $a_{i,i+1}$ is the arc going out of n_i and entering n_{i+1}, $i = 1, \ldots, k-1$. As before, the length of a directed walk equals, by definition, the sum of the lengths of its arcs. One of the graph optimization problems closest to our intuition and to our daily experience is the problem of finding the shortest walks from node s to all other nodes. However, in the case where the graph contains negative-length arcs some clarifications have to be made. As customary, let us begin by looking at our example (Fig. 8.9). All the nodes A, B, C are reachable from s, respectively, through the arcs $s \to A$, $s \to B$, $s \to C$ of length 0. However, the presence of negative-length arcs produces a particular phenomenon. Indeed, suppose we reach node B passing through the arc $s \to A$ and the arc $A \to B$ with a path of length -126. In doing so, we have reduced the minimum distance from s to B, which is now -126. Then, passing again through A by the arc $B \to A$ of length 44, also the minimum length of the walk from s to A would be updated to $-126 + 44 = -82$. In fact, each time that we go through the negative cycle $A \to B \to A$, we reduce the distance between s and nodes A and B. At the limit, going along this cycle infinite times, we could reduce such a distance indefinitely. This phenomenon occurs if and only if the graph G contains a cycle of negative length (reachable from s). Now, suppose that TP3 is removed from Z. The modified difference constraints graph is represented in Fig. 8.10.

Observe that negative cycles do not exist any more, allowing us to compute the minimum (finite) distances from s to all other nodes. In fact, A is reachable by the zero-length arc $s \to A$; B through the walk $s \to A \to B$ of length -126; C through the walk $s \to A \to C$ of length -186. It is indeed possible to prove that such distances provide a feasible solution to the system of inequalities of Table 8.3. In general:

Fig. 8.10 The difference constraints graph after TP3 removal

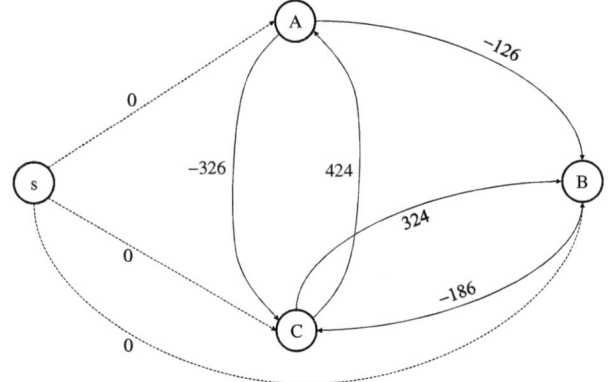

> *Property 3.* Let $G = (N, A)$ be a graph associated with a system of difference constraints. If G does not contain negative length cycles, a solution to the system is obtained by setting t_i equal to the length of the shortest path from s to node i, for each node $i \in N$.

The proof of this property requires advanced knowledge of optimization theory and goes beyond our purposes.

We also observe that the solution may contain negative t_i values. However, through a *scaling* operation, that is, by summing to all values the absolute value of the shortest distance (in our case that of the walk $s \to A \to C$), a new solution of the system is obtained with non-negative values, thus representing feasible transmission delays. For example, in our case, this yields $t_A = 186, t_B = 60$ and $t_C = 0$, which guarantee the coverage of the TP set $S = \{1, 2\}$, as shown in Fig. 8.8.

In conclusion, Step 3.b of Algorithm in Table 8.2 is reduced to the computation of shortest paths from s to the other nodes of G. It is interesting to note that, by simple shrewdness, the algorithm that finds the shortest walks can also be exploited to implement Step 2. Intuitively, if in the graph there is a negative cycle, the algorithm is "attracted" to it, in the sense that it will tend to run it many times, each time reducing the lengths of the paths from s to the nodes of the cycle, as evidenced in the example. Then a simple check can be implemented that, in the presence of this phenomenon, stops the algorithm and returns the cycle identified. In fact, the entire algorithm in Table 8.2 is nothing but a sequence of shortest walk computations!

8.5 From Shortest Walk to Television

Retracing the discussion, we can observe that, starting from physical elements and technological aspects of the system, we came to the definition of alternative mathematical representations of the problem, such as the system of difference constraints, or the associated graph. This modeling process shows deep connections between apparently very different problems, such as the design of a television network and the search for the shortest walk in a generic network (e.g., a road network). Besides the cultural interest and curiosity stimulated by a similar connection, the modeling process has practical significance: new representations of the planning problem allow for a better identification and formalization of solution algorithms. In other words, the choice of the algorithm was carried out by reducing the real problem to general paradigms of optimization, such as the search for shortest walks on graphs, for which efficient algorithms are available.

It is worthwhile to mention that advanced developments of the methodologies described in this chapter have already had a significant practical impact on the television that we see every day, especially with regard to terrestrial networks in digital technology that are currently being implemented.

8.6 Bibliographic Notes

A comprehensive survey of algorithms for finding shortest walks in networks along with other optimization problems concerning "Flows on Networks" can be found in the book by Ahuja, Magnanti and Orlin [4]. A simple and complete introduction to wireless communication systems is given by Rappaport [94]. Besides technological aspects, the book also illustrates models and techniques for radio network planning.

The algorithms for coverage evaluation are recommended by international bodies, such as the *International Telecommunication Union* (ITU, www.itu.int) and the *European Broadcasting Union* (EBU, www.ebu.ch). In particular, EBU defined the standards for service evaluation in digital television [36].

The *White Book on Terrestrial Digital Television* [2] was produced by the Italian Authority for Communications (AGCOM). This work involved all the main stakeholders and represented the first step towards the transition from analogue to digital broadcasting networks in Italy. Specifically, it started the debate about the possible configuration for digital multiplex and the need for optimization in network planning and spectrum management. The *White Book on Terrestrial Digital Television* was also the methodological background for a regulation law of AGCOM known as the *National Plan of Digital Radiotelevision Frequencies* [1].

Chapter 9
Algorithms for Auctions and Games

Vincenzo Bonifaci and Stefano Leonardi

Abstract Economics is one of the diverse aspects of our life in which algorithms play a – perhaps subtle – role. Whether we are buying an item through an eBay auction, or choosing a pricing scheme from our telephone company, we are participants in distributed decision-making processes having consequences that may affect our personal "welfare". And with the advent of the Internet and the Web, there is an increasing demand for algorithms that run, manage or analyze such economic transactions. In this chapter, we discuss some of the basic notions that underlie an algorithmic view of economic and strategic interactions.

9.1 Introduction

The exchange of ideas between the theory of algorithms and the economic theory of games is one of the most fascinating facets of computer science in the last decade. Such an encounter originated from the development of the Internet and the World-Wide Web (see Chap. 5) as organic, self-organizing systems, whose evolution is guided by a multitude of independent agents that operate according to economic principles. Computer science has taken inspiration from the economic theory of games in order to understand the economic mechanisms that motivate the agents of the network; in fact, the strategic interaction between agents is now considered an essential aspect of algorithm design in such contexts.

V. Bonifaci (✉)
Istituto di Analisi dei Sistemi ed Informatica "Antonio Ruberti",
Consiglio Nazionale delle Ricerche, viale Manzoni 30, 00185 Roma, Italy
e-mail: vincenzo.bonifaci@iasi.cnr.it

S. Leonardi
Dipartimento di Ingegneria Informatica, Automatica e Gestionale,
Sapienza Università di Roma, via Ariosto 25, 00185 Roma, Italy
e-mail: leon@dis.uniroma1.it

The Internet is a decentralized system in which agents are independent and able to take individual decisions. In such scenarios it is important to foresee the states that the system is going to enter, since it is normally impossible to enforce some particular, favorable state. Consider, for example, the selection by the network agents of the path and transmission speed when transferring data from some origin to some destination. The strategy of each agent can consist in selecting, for each transmission, the path with the smallest estimated delay. Then we may, for example, ask whether the system will reach a state in which congestion is much larger than that of a state in which routing strategies are centrally planned. Or whether the system can oscillate from state to state due to the continual change of the strategies of the agents. The interaction between agents makes the adoption of the points of view of the theory of games unavoidable, and in particular calls for the notion of equilibrium states, those states of the system from which no agent has an incentive to deviate.

In recent years, several methodological contributions have been provided by algorithmics to economics and to game theory in particular. First of all, we remark that the theory of algorithms, mathematical optimization and the theory of computational complexity are tackling the delicate questions of the existence of equilibria and of the hardness of computing them. Secondly, a study has begun of the inefficiency of the solutions determined by the equilibrium states, in other words a quantitative comparison between the equilibrium solutions and the best solutions that could be imposed by a centralized "enlightened dictatorship". Finally, given a view of the Internet as an open and democratic system aiming to guide the agents towards solutions that are more favorable towards the overall community, algorithms and protocols have been proposed that include incentives and penalties.

Computer science and network science also bring an important contribution to economic sciences, since the ever larger and pervasive diffusion of the Internet determined the migration of several economic and commercial activities on the net and created more of them, including ones that were unthinkable before, things like electronic commerce and online computerized auctions with a multitude of participants (eBay, Amazon, etc.). This determines the need to design algorithms and software that manage the commercial transactions electronically. Another important aspect has been the advent of digital goods in unlimited supply and the availability of commercial spaces on digital media such as the Web, blogs, forums and social networks.

We would like to stress one more aspect of the encounter between computer science and economics. The Internet and the World-Wide Web can be viewed as formed and governed by the action of a multitude of autonomous, rational agents that operate with the goal of optimizing their own individual "payoff functions". Network algorithms and protocols have to operate on data supplied by these agents. But such data could be manipulated by the agents in order to optimize their own utility. For example, an algorithm searching for shortest routes along the Internet will have to acquire as input the costs for traversing (sub)networks, costs that are estimated and declared by autonomous systems, the independent economic entities that manage the networks. The traditional approach to algorithm design should then be rebuilt to accommodate the issue of input data manipulation from the part of rational agents. The fundamental question there is whether it is possible to design algorithms that efficiently solve some algorithmic problem and at the same time give incentives to the agents to reliably report that part of the input data that is their private information. *Mechanism design* offers in this direction an essential methodological reference and an important contribution from economic sciences to computer science.

The arguments that we discussed should have convinced the readers that the encounter between algorithms and economic mechanisms is a necessity imposed by the Internet, an information system that in the last two decades revolutionized the modes of production and communication. The fact remains that the fundamental principles and the quality properties that are required by algorithms and economic mechanisms are deeply different, and it is not at all clear that for one of the two disciplines it should be possible to relax them while adopting those from the other discipline. In the case of algorithms, such requisites are usually expressed in terms of the computational resources necessary for execution or in terms of the quality of the computed solution. In the case of economic mechanisms, these requisites are expressed in terms of objectives such as the equilibrium between demand and supply, the maximization of the agents' profit or utility and, finally, the impartiality of a recognized authority with respect to all the participants of some economic game. Such an integration requires, then, a deep knowledge of both fields and the development of new models, algorithms and mechanisms, as well as methods for their analysis.

In this chapter we intend to present some fundamental concepts of the theory of games and some examples that illustrate how algorithms inspired by the theory of games are essential for many economic activities that happen daily on the Internet.

9.2 Games and Solution Concepts

9.2.1 *Prisoner's Dilemma*

To better understand the approach of game theory to the analysis of conflict situations, consider the following famous scenario, proposed in 1950 by American mathematicians Merrill Flood, Melvin Drescher and Albert W. Tucker. Two crime suspects, R. and C., have been captured by the police and have been charged with a major offense. In order to acquire the necessary evidence, the prosecutor is trying to separately persuade each of the suspects to confess his crime. The suspects are held in separate cells, and no communication can occur between them. The prosecutor proposes the following to each: if the prisoner confesses, his jail sentence will be cut in half – but only if his accomplice does not also confess, otherwise there would be no choice but to convict both. If the prisoner does not confess, he will get the maximum possible time in jail if the prosecutor is successful; otherwise, both suspects will be convicted anyway for some other minor offense.

Table 9.1 Prisoner's Dilemma

	Confess	Silent
Confess	−5　　−5	−10　　0
Silent	0　　−10	−1　　−1

We reach the situation summarized in Table 9.1. The action selected by R. determines the row of the table, the one selected by C. the column; this is why R. is also called the Row player and C. the Column player. In each of the four entries of the table, the bottom left number denotes the payoff (also called *utility*) of R. and the top right number the payoff of C. Most numbers are negative since they denote the penalty due to the number of years to be spent in prison: serving 10 years is worse than serving 5. A zero means that the criminal was let free.

What will each player do? Imagine taking R.'s role. Observe that if C. confesses, it is better to confess, since you'll have to serve 5 years instead of 10. And if C. doesn't confess, it is better to confess too: you'd be let free instead of spending a year in jail. So no matter what C. does, confessing is a better option than not confessing! An action with such a property – the property of being preferable to all other actions, independently of the other players' choices – is called a dominant strategy.

Looking at the table, one sees that the roles of R. and C. are symmetrical, therefore confessing is a dominant strategy for C. too. Thus, the overall outcome is that both prisoners will have to spend 5 years in jail, more than each of them would have if they both chose not to confess. This result may sound paradoxical, but several experiments in psychology and sociology show that outcomes of this kind are actually not uncommon in practice.

It should be intuitively clear that the Prisoner's Dilemma is not limited to the simple scenario outlined above. It has been applied in the most diverse contexts and several examples can be found also in telecommunication and computer networks. A more general situation is the setup and usage of some public good. Say that a public good is worth 3€ to each of the two players, Row and Column. Some investment is necessary to create the good, so the good will be available only if 4€ total are invested. Each player may volunteer to invest or not: if he is the only volunteer, he will pay 4€, otherwise the cost is shared equally. The payoff of a player is the value of the good (or zero, if the good is not created), minus the investment cost. The resulting payoff matrix is given in Table 9.2.

The situation of each player is similar to that in the Prisoner's Dilemma: there are two choices, one of which is better from an opportunistic point of view, but that damages the other player. Although numerically different, the payoff table has in fact the same structural properties of the payoff table for the Prisoner's Dilemma: namely that the more aggressive option is, for both players, a dominant strategy. Therefore, the theory predicts that no investment will be made.

Table 9.2 Provision of a public good

	Free ride	Invest
Free ride	0 0	−1 3
Invest	3 −1	1 1

Table 9.3 Battle of the Sexes

	Movies	Concert
Movies	4 3	1 1
Concert	2 2	3 4

A final remark on the pessimistic outcome that game theory predicts in the Prisoner's Dilemma. The result is based on several assumptions that are not necessarily realistic: that the players are rational; that they care exclusively about their own good; and that the game is played exactly once. In the absence of one or more of these conditions, it is possible to have outcomes that are different and more optimistic. In fact, Merrill Flood, one of the discoverers of the Prisoner's Dilemma, once made the following proposal to a secretary of the institute he was working in: she could either immediately accept 100 dollars, or be given 150, but in the second case she would need to find an agreement on how to split the money with a second secretary that had been kept unaware of the proposal. The secretary chose the second option and agreed with her colleague to get 75 dollars each, even though she could easily have kept the 100 dollars without her colleague ever getting to know anything.

9.2.2 Coordination Games

In some strategic situations, conflicts may arise simply due to the impossibility for the players of the game to coordinate their choices. Consider another scenario, known in game theory as the "Battle of the Sexes". A man and a woman are deciding how to spend the evening out: the man would like to go to the movies, while the woman would prefer a concert. It is too late to consult and each of them should reach the chosen location directly from his or her office. Both of them prefer to spend the evening together rather than alone, but, given the choice, the man would prefer the movie while the woman would prefer the concert. A possible payoff table for the game is reported in Table 9.3, where the values denote again the utility of each action combination.

What does game theory predict in such a case? The answer is less clear-cut than that we saw for the Prisoner's Dilemma. If the woman knew for certain that the

Table 9.4 Rock–Paper–Scissors

	Rock	Paper	Scissors
Rock	0	−1	1
Paper	1	0	−1
Scissors	−1	1	0

man would go to the movies, she would choose the movies over the concert; if she knew that the man would go to the concert, she would be happy to join him there. In other words the best response to the man's action of going to the movies is for the woman to go to the movies, and the best response to the man's action of going to the concert is for the woman to go to the concert. A similar reasoning holds of course for the man. The two solutions (Movies, Movies) and (Concert, Concert) are therefore "stable" in the sense that in each of them the action chosen by each player constitutes a best response to the other player's choice.

Such "mutual best response" outcomes are called Nash equilibria, from the name of the American mathematician and Nobel prize for Economics winner John F. Nash, who introduced and studied them. In the Battle of the Sexes, the only equilibria are those in which the woman and the man spend the evening together: for example, the outcome in which the woman goes to the concert and the man goes to the movies is not a Nash equilibrium, since the man could improve his payoff by "changing his mind" and going to the concert instead.

A small weakness of the Nash equilibrium as a solution concept of a game is that it is not necessarily unique, as can be seen from the example of the Battle of Sexes, in which there are in fact two Nash equilibria. On the other hand, this ambiguity captures the uncertainty of outcomes in a conflict situation.

9.2.3 Randomized Strategies

The game known as Rock–Paper–Scissors is a simple and popular two player game. The players simultaneously represent, by a gesture of their hand, an object among paper, scissors and rock. Rock beats scissors, scissors beats paper, and paper beats rock. If the players choose the same object, the game is tied. A player gets payoff 1 if she wins, −1 if she loses, and 0 if there is a tie. Therefore, the sum of the two payoffs of the players, whatever the outcome of the game, is always zero; a player can win only as much as her opponent loses – something that was not the case with other games such as the Prisoner's Dilemma or the Battle of the Sexes. Such a game is called *zero-sum*. In a two-player zero-sum game it is sufficient to specify the payoff of one of the two players, say the Row player, as in Table 9.4.

The Rock–Paper–Scissors game has an interesting property: it admits no Nash equilibria. The reason is that the best response of a player to a given action of the opponent renders the action of the opponent a "bad response": if we knew our opponent would play paper, we would play scissors, but then our opponent would play rock, etc. Therefore there is no Nash equilibrium in the sense we previously discussed.

What can game theory say about such cases? The key idea lies in generalizing the notion of strategy to allow probabilistic or *mixed* strategies (see the box "Events, Probabilities and Expected Values"). An example of mixed strategy is: play paper with probability 70%, scissors with probability 20% and rock with probability 10%.

As a consequence of this generalization, the concept of payoff of a player is substituted by that of expected payoff, which is simply the average utility obtained by choosing the actions according to the prescribed probabilities. If, for example, we play 70% paper, 20% scissors, 10% rock, and our adversary answers with 100% scissors, our expected utility will be

$$0.7 \times (-1) + 0.2 \times 0 + 0.1 \times (+1) = -0.6.$$

To remark on the difference with such mixed strategies, the strategies in which each player selects a single action are called pure strategies. A pure strategy can be seen as a very special case of mixed strategy.

Events, Probabilities and Expected Values

Mathematically, the probability of an event A is a real number between 0 and 1, often denoted as $\mathbf{Pr}[A]$. When A is an impossible event, $\mathbf{Pr}[A] = 0$, while when A is certain, $\mathbf{Pr}[A] = 1$.

The *complement* event to A (which occurs if and only if A does not occur) has probability $1 - \mathbf{Pr}[A]$.

When A and B are *independent* events, the probability that they simultaneously occur is equal to $\mathbf{Pr}[A] \cdot \mathbf{Pr}[B]$. For example, the probability of obtaining two heads when tossing two coins is equal to $1/2 \times 1/2 = 1/4$.

When A and B are *mutually exclusive* events, the probability that at least one of them holds is equal to $\mathbf{Pr}[A] + \mathbf{Pr}[B]$. For example, the probability to obtain a 5 or a 6 after throwing a standard six-faced die is equal to $1/6 + 1/6 = 1/3$.

Finally, when a numerical variable X can take the values $x_1, x_2, \ldots, x_n$, with probability $p_1, p_2, \ldots, p_n$, respectively, its *expected value* is given by the formula

$$p_1 \cdot x_1 + p_2 \cdot x_2 + \ldots + p_n \cdot x_n.$$

The conceptual leap from pure strategies to mixed strategies has an interesting and somewhat unexpected consequence: any game that is finite, in other words any game that has a finite number of players and a finite number of strategies, always admits at least one mixed strategy equilibrium. Such a fundamental property was proved by John Nash in 1950 and has since been known as Nash's theorem.

What is, then, a Nash equilibrium in mixed strategies for the Rock–Paper–Scissors game? Assume we play the three actions at random, each with probability $1/3$. If our opponent plays paper with probability p, scissors with probability s, and rock with probability r, his expected utility will be

$$(1/3) \times 0 \times p + (1/3) \times 1 \times p + (1/3) \times (-1) \times p$$
$$+ (1/3) \times (-1) \times s + (1/3) \times 0 \times s + (1/3) \times 1 \times s$$
$$+ (1/3) \times 1 \times r + (1/3) \times (-1) \times r + (1/3) \times 0 \times r$$
$$= 0.$$

Therefore the expected payoff of our opponent will be zero, independently from the mixed strategy that he chooses. If he, too, is choosing with the same probabilities rock, paper and scissors, then the expected payoff of both players will be zero and neither of them will have a way to obtain an expected payoff larger than zero by adapting their strategy. In other words, we will have a situation of mutual best response, exactly as required by the definition of Nash equilibrium. In fact, in this case the outcome we described is the only Nash equilibrium of the game, although this uniqueness property does not hold in general, as we saw in the Battle of the Sexes example.

The reader may wonder whether actual Rock–Paper–Scissors games are played as the theory prescribes. It is certainly true that if one follows the optimal strategy, then it is impossible to achieve a negative expected utility (although it is always possible to be particularly unlucky!). However, many "real-world" players, due to limitations of several kinds, such as not being able to choose the actions in a perfectly random fashion, may not be playing an optimal strategy but some slightly different mixed strategy. In this case we might obtain an expected payoff larger than zero by adopting a mixed strategy different than the perfectly uniform one. For example, if our opponent plays scissors and paper with the same probability, but never plays rock, then we can increase our expected utility by always playing scissors. Our expected payoff then becomes positive, since

$$1 \times 0.5 \times 0 + 1 \times 0.5 \times 1 = 0.5.$$

For this reason, in games such as Rock–Paper–Scissors, the players that more often succeed are those that try to learn the adversary's strategy, at the same time adapting their own strategy to it.

9.2.4 Hawks and Doves

So far, we exemplified potential conflict situations through some simple but paradigmatic two-player games: the Prisoner's Dilemma, the Battle of the Sexes

9 Algorithms for Auctions and Games

Table 9.5 Chicken

	Swerve	Keep
Swerve	0 0	1 −1
Keep	−1 1	−100 −100

and Rock–Paper–Scissors. We add an equally interesting game to the list, called "Chicken". The scenario is as follows: two contestants start driving cars at each other at high speed. The first of the two that, by swerving, deviates from the collision trajectory will be the loser, in other words the "chicken". The player that resists swerving until the end will instead be the winner. Clearly, a participant in Chicken should try to resist as much as possible in order to overcome his opponent, but if both players act out this idea, the result is a disaster!

Analyzing the game (Table 9.5) in light of the concepts discussed so far, we observe that there are only two Nash equilibria in pure strategies: the one in which the first player swerves and the second keeps the trajectory, and the one in which the opposite happens. Is there by any chance some Nash equilibrium in which the strategies are all probabilistic? Indeed there is. Denote by R_s and R_k the probability with which the first player (the Row player) swerves or keeps the trajectory, respectively. If we consider the configuration in which $R_s = 99\%$ and $R_k = 1\%$, then we notice that

$$0 \times R_s + (-1) \times R_k = 1 \times R_s + (-100) \times R_k = -1.$$

This implies that the expected payoff for the second player is the same (-1) independently of the action he chooses. Similarly, if $C_s = 99\%$ and $C_k = 1\%$ are the probabilities with which the second player swerves or keeps the trajectory respectively, we have

$$0 \times C_s + (-1) \times C_k = 1 \times C_s + (-100) \times C_k,$$

so that also the expected payoff of the first player is the same independently of his action. Such a combination of mixed strategies therefore constitutes a Nash equilibrium.

Games akin to Chicken are those in which the players have to decide whether they want to be aggressive "hawks" or peaceful "doves". The Cuban missile crisis of 1962 was analyzed by the advisors of US President J.F. Kennedy and classified exactly as a situation of this kind, where the catastrophic outcome could have been a nuclear war. Kennedy decided to let USSR Chairman N. Khrushchev know that the United States would not have played the dove's role, even to the point of having to enter the war. Luckily for everybody, Khrushchev chose to concede and be a "dove".

The Chicken game also allows us to introduce an important extension of the notion of Nash equilibrium, proposed by the Israeli mathematician Robert

J. Aumann (also a Nobel prize for Economics) in 1974. In this extension, known as correlated equilibrium, we consider acceptable equilibria all outcomes of the game in which players do not have an incentive to change their own strategy, assuming that the strategy that the players follow is suggested in some way by a trusted third party. For example, consider a third party that with probability 1/2 suggests to the first player to swerve and to the second player to stay on track, and with probability 1/2 gives the opposite suggestions. In this case, if one of the two players assumes that the other one behaves according to the suggestion of the third party, he will have, in fact, an incentive to behave exactly as suggested; in other words, the suggestion is self-enforcing. This is not at all a phenomenon that is far removed from daily experience: a traffic light is nothing but such a trusted third party in the Chicken-type game that any two car drivers play when they have to go through the same crossing from different directions.

9.3 Computational Aspects of Game Theory

We have seen through the previous example how rigorous theorems such as Nash's theorem ensure, for several games, the existence of the equilibria we are interested in. We can ask ourselves whether these equilibria can actually be computed via some algorithm: knowing that an optimal strategy for a given game exists does not help us much if we do not know how to determine it. For this reason the computational aspect of game theory is crucial. We may even go further and say that if a given type of equilibrium is hard to compute, then most likely that type of equilibrium does not fully capture the realistic outcomes of a game, since after all the participants in a game, be they individuals or companies or specifically designed software, always have a limited computational power. The computational properties of equilibria therefore help us understand what equilibrium concepts are indeed more realistic.

9.3.1 Zero-Sum Games and Linear Optimization

In the case of games where players move simultaneously, the computation of equilibria is more or less complicated depending on the type of equilibrium that is sought. For example, in the case in which one seeks a Nash equilibrium in pure strategies, it is indeed sufficient to consider all combinations of the players' strategies and for each combination verify whether both players are playing a best response action. If the first player has m actions available and the second player has n, we have to consider $m \times n$ cases, and for each of those we must compare the payoff with other $m + n - 2$ values of alternative payoffs. As we have already seen, though, an equilibrium in pure strategies might not exist.

More interesting is the case of Nash equilibria in mixed strategies. If we consider arbitrary games (even with only two players) things get complicated pretty quickly.

Table 9.6 An example of a zero-sum game

	C	D
A	2	−1
B	1	3

We thus start from the relatively simpler case of zero-sum games; this case was analyzed for the first time in 1928 by John von Neumann. As we have seen with the Rock–Paper–Scissors game, zero-sum games can be described by a single matrix in which the entries indicate at the same time the payoff for the first player and the cost for the second player, as in Table 9.6 where, for example, if the first player chooses strategy A and the second strategy C, then the first player gains 2 and the second pays 2.

The basic idea behind von Neumann's result is that an equilibrium strategy should be stable even when it has been revealed to one's opponent. If the mixed strategy of the Row player is known, the Column player will select a mixed strategy that minimizes the payoff of the Row player. By foreseeing this, the Row should then select a strategy that minimizes his own maximum loss, or in other words that maximizes his minimum gain. Such a strategy his called a *minimax strategy*: some mathematical details are given in the box "Minimax Strategy". The main point is that in this case the equilibrium strategies can be computed efficiently.

Minimax Strategy

A minimax strategy can be expressed by a system of inequalities, where the variables a and b represent the probability with which the player chooses his strategy A or B, respectively, and where variable v represents the gain obtained by the player.

$$\max \ v$$
$$2a + b \geq v$$
$$-a + 3b \geq v$$
$$a + b = 1$$
$$a, b \geq 0.$$

The first two conditions ensure that no matter the answer of the Column player, the payoff of the Row player will be at least v. The other conditions simply ensure that a solution of the system represents a mixed strategy. By solving the system – for example graphically (see Fig. 9.1) – we obtain $a = 2/5$, $b = 3/5$, that is to say that the Row player should play action A with probability 40 % and action B with probability 60 %. That ensures an expected payoff $v = 7/5$ to the Row player.

(continued)

Fig. 9.1 Equilibrium computation for a simple zero-sum two-player game

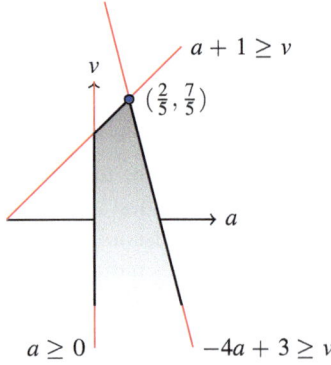

(continued)

Clearly, a similar argument can work for the Column player too. In this case, if w represents the minimum loss by the Column player, we get $w = 7/5$. The fact that v equals w is a general property that is ensured by von Neumann's theorem: when both players follow a minimax strategy, the resulting outcome is an equilibrium. In general such systems of inequalities constitute instances of linear optimization problems, a well-studied field in mathematical optimization. Algorithms are known that solve such problems in polynomial time (see Chap. 2). The same algorithms can thus be directly applied to the computation of mixed strategy equilibria in zero-sum games with two players.

9.3.2 Fixed-Points: Nash's Theorem and Sperner's Lemma

What happens when we move from zero-sum games to the general case? Is it still possible to efficiently compute the Nash equilibria in mixed strategies? It turns out that no one has a definitive answer to such an apparently simple question.

A natural approach, when trying to understand how to find a Nash equilibrium in a non-zero sum game, is to take a step back and analyze the proof of Nash's theorem – in other words, to understand why an equilibrium in mixed strategies must exist. We will not give here the details of the proof, but we will hint at some of the ideas on which it is based.

Nash's theorem can be seen as a so-called fixed point theorem, that for a given function F asserts that, under suitable conditions, the equation $F(x) = x$ always admits a solution – in other words, that the function F has at least one fixed point.

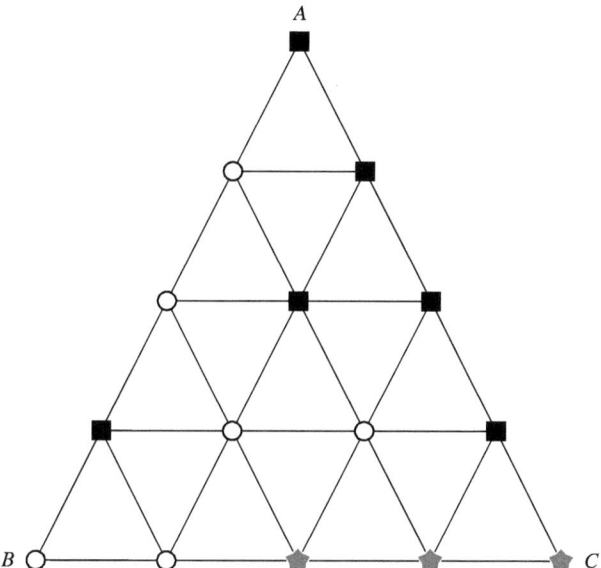

Fig. 9.2 A valid coloring of a subdivided triangle ABC

In the case of Nash's theorem, x represents a list of mixed strategies (one for each player), and the function $F(x)$ gives a new list of mixed strategies in which every player is adopting a best response to the configuration in which the other players choose their actions according to what is specified by x. The equation thus represents the fact that we seek a list of mixed strategies such that each is a best response to the other ones.

To illustrate one of the ideas on which the theorem relies, we will discuss a result used in the proof of the theorem and strongly linked to it, known as Sperner's lemma. We will not formulate the lemma in its full generality. However, one of the consequences of this lemma is the following: take some triangle ABC and arbitrarily subdivide it into smaller triangles, as in Fig. 9.2. Then color the vertices of all small triangles with one of three colors (say white, black and gray), while obeying the following rules:

1. The three vertices A, B, C of the original triangle should be colored with three different colors;
2. The vertices that lie on a same edge of the original triangle should not be colored with the color assigned to the vertex opposite to that side; for example, vertices lying on the AB line cannot be colored with C's color;
3. The vertices in the interior can be arbitrarily colored.

Sperner's lemma asserts that, independently of how we color the vertices, there will always exist a triangle of the subdivision whose vertices have three distinct colors. And, in fact, in the case of Fig. 9.3 such a triangle exists. Why is that?

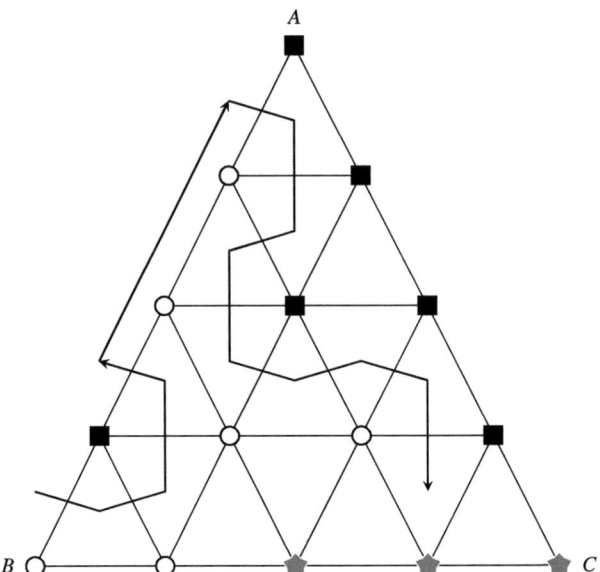

Fig. 9.3 Illustration of the proof of Sperner's lemma

To understand, mentally picture each triangle of the subdivision as a triangular room with three walls, one per side. Each wall has a door in it, if the corresponding side of the triangle has one white vertex and one black vertex. It is easy to check that such rooms may have zero, one or two doors, but never three. Moreover, if a room has a single door, it must correspond to a triangle with three differently colored vertices, because if that was not the case there should be an additional white–black side, and the doors would be two.

Observe that from the outside of the original triangle ABC there is an odd number of accessible doors (again, this can be seen to be always true). Take any one of this doors and follow the path through each door. Notice that we never have a choice, since the number of doors in a room on the path is one or two. The path will necessarily end either in a room with a single door, or again outside the ABC triangle. But in this last case the path has used two doors facing the outside of the triangle, and since there was an odd number of those doors, there must be another door from which we can proceed in a similar fashion. Thus the only way to end this process is to find a room with a single door, corresponding to the triangle with three distinct colors whose existence was claimed by the lemma. Figure 9.3 illustrates the argument.

The above argument may perhaps give a vague intuition of why finding Nash equilibria in mixed strategies appears to be difficult. The procedure to determine the triangle that verifies Sperner's lemma is correct, but in general it may require very long paths, which in turn correspond to a long computation time. Nothing, however, prevents that in principle one could devise a smarter, more "direct" method of determining the correct triangle, even though that appears unlikely.

9.3.3 Mixed Nash Equilibria in Non-zero-Sum Games

There is an algorithm, although markedly inefficient, that allows us to determine a Nash equilibrium in mixed strategies for any two-player game. An idea at the basis of the algorithm is the notion of *support* of a mixed strategy. The support is simply the set of actions that in the mixed strategy considered are selected with probability larger than zero. In other words, it is the set of pure strategies that concur to form a given mixed strategy. Now, a property of equilibria is that a mixed strategy constitutes a best response if and only if all the pure strategies in its support are best responses. This fact is useful to determine a Nash equilibrium. In the box "Equilibrium Computation" a complete example is given, with all the corresponding calculations.

Equilibrium Computation

Consider the following example.

	C	D
A	1	3
	2	0
Silent	2	1
	1	4

The pure strategies of the Row player are A and B, those of the Column player C and D. Let us ponder whether a Nash equilibrium exists in which the support of the first player is $\{A, B\}$ and that of the second player is $\{C, D\}$. Call a, b, c, d the probabilities assigned to the respective strategies. Then it should hold

$$a, b, c, d > 0,$$
$$a + b = 1,$$
$$c + d = 1.$$

Moreover, in order for (a, b) to be a best response to (c, d), and since we are assuming that both A and B are in the support of (a, b), A must be a best response to (c, d) and should thus give an expected payoff at least as large as that would be obtained with B:

$$2c \geq c + 4d,$$

(continued)

> **(continued)**
>
> but by a similar argument B should be a best response to (c, d):
>
> $$c + 4d \geq 2c.$$
>
> We thus obtain $2c = c + 4d$, that combined with the previous equations gives $c = 4/5$, $d = 1/5$. Analogously we can deduce, from the fact that both C and D should be a best response to (a, b), that $a + 2b = 3a + b$ and so $a = 2/3$, $b = 1/3$. Such an argument shows that the main difficulty in finding a Nash equilibrium is in the determination of the support sets of the two players. Once the supports are known, as we explained it is sufficient to verify that a given system of linear inequalities admits a solution. A possible algorithm to determine a Nash equilibrium then, however inefficient, consists in enumerating all possible pairs of supports and for each of them verifying if it gives rise to an equilibrium by solving the linear system. The running time of such an algorithm is dominated by the number of possible pairs of supports, that in the case of m actions for one player and n for the other one, is roughly 2^{m+n}.

9.4 Inefficiencies

9.4.1 The Tragedy of the Commons

In many of the examples discussed so far, we have seen how one can mathematically formalize the kind of individual behavior that emerges from the interaction of rational agents, each of whom is driven to his own goals. A natural question is: what happens to the overall system? What levels of "social welfare" are obtained when each agent is pursuing his own goal separately? For such questions to make sense we need to define what we mean by social welfare. There are many possible definitions that are equally valid. For example, with social welfare we could mean, in a utilitarian perspective, the sum of the payoff of all the players; or the payoff of the player with the smallest payoff (the "poorest"). In any case it should intuitively be clear that when the users of a system individually pursue their own goals, the resulting social welfare is not necessarily maximized. In other words, the solution determined by the agents will not in general be a globally optimal solution.

Consider, for example, a simple scenario in which a set of 100 users shares the access to the same Internet connection, by means of a certain finite transmission bandwidth B. Every user can regulate the amount of bandwidth that he intends to use. The strategy of the ith user consists in the fraction $x(i)$, with $0 \leq x(i) \leq 1$, of

9 Algorithms for Auctions and Games

the amount of bandwidth used (so that the user gets a bandwidth equal to $x(i) \cdot B$). Assume that the payoff of a user depends on the amount of bandwidth he consumes, but also on the amount of bandwidth left unused by the other players, according to the formula

$$u(i) = x(i) \times (1 - x(1) - x(2) - x(3) - \ldots - x(100)).$$

The second term in this formula captures the fact that the latency of the communication channel is lower when the channel is less congested, while if the channel is almost saturated the latency is very high and thus the payoff quickly drops. (For simplicity we do not require that the sum of the $x(i)$ fractions be inferior to 1, so the second term can even become negative.) We thus obtain a multiple player game. Such a game is not finite – the number of players is finite, but the strategies are values from a continuous set and are thus infinite – so we cannot invoke Nash's theorem directly.

Still, in this case a Nash equilibrium exists; we can find out that it corresponds to the solution in which each user has a fraction of the bandwidth equal to $1/101$. The communication channel in this solution is almost completely saturated, since it is used for a total fraction equal to $100/101$. The payoff of each user will then be $1/101 \times (1 - 100/101) = 1/(101)^2$. How large is the social welfare in this case? If by social welfare we mean the total utility of the players, we obtain a social welfare equal to $100/(101)^2$, roughly 0.01. However, if we could have forced each user to use a bandwidth fraction equal to $1/200$, half of the total bandwidth would have stayed unused, so that the payoff of each user would have been equal to $1/400$, and the total payoff equal to $1/4 = 0.25$. Such value is 25 times larger than the value obtained at the equilibrium (0.01), so that we can say that in this case the independence of the players caused a large decrease of the social welfare. Such a negative phenomenon is well-known in economics under the name of *tragedy of the commons*. It manifests itself every time that the individual interests of a group of users tend to destroy the advantages deriving from the use of a common resource. Clearly, not all economical interactions are of this kind and sometimes the individual interest gets close to the collective one, but such 'tragedies' are indeed frequent.

Although such phenomena were known since long ago, only more recently have researchers from the computer science community, starting with Elias Koutsoupias and Christos Papadimitriou, analyzed them in a deeper quantitative way, through the notion of *price of anarchy*: that is, the ratio between the globally optimal social welfare and the social welfare arising from the equilibrium (or between the social cost at the equilibrium and the optimal social cost). The closer such ratio is to one, the more reasonable it is to claim that the individual interest approximately coincides with the collective interest, while if the price of anarchy is very high – as in the case of the bandwidth sharing game – the outcomes in the two settings can be very different. The notion of a price of anarchy has been applied and continues to be applied in the study of many economical scenarios, in particular those related to networks.

Fig. 9.4 Pigou's example

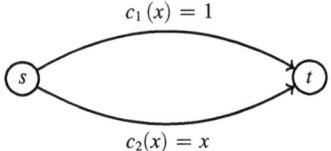

9.4.2 Routing Games

Imagine a familiar scenario: we need to move by car from one end of the city to the other, and the choice of our itinerary will have deep impact on the travel time. Such time will also depend on how many other car drivers will choose the same itinerary, due to possible congestion effects. We thus see that the scenario can be modeled as a game in which the players are the drivers and the actions are the itineraries. What are the equilibria of the game, and how inefficient are they? The question is not futile, also because a similar model can be applied to computer networks, with the difference that the flows to be routed are flows of data instead of flows of vehicles. It should not be surprising then that the answer has been in part given by computer scientists, as well as by traffic engineers.

To understand it we first have to better specify our model. We represent the traffic network as a graph in which nodes are the locations of interest and the arcs represent the connections from one place to the others. Moreover, to describe the possible congestion effects, we assign to each arc a cost (or "latency") function that gives the travel cost incurred for each level of traffic on that particular connection. For example, in the graph of Fig. 9.4 the upper arc has a constant cost function (function c_1), while for the lower arc the cost function is the identity (function c_2).

We finally select an origin point s, a destination point t and a flow value, that is, an amount of traffic to be routed from s to t. We can imagine that such flow is composed of an infinite number of "traffic particles", each of which has no influence by itself, but that can behave independently (in this formulation we thus have a game with infinitely many players – the existence of an equilibrium is not ensured by Nash's theorem, but can in fact be proved by similar means). The overall cost of a certain flow is obtained by summing, on each arc, the product between the quantity of flow traversing the arc and the latency of the arc. For example, if in Fig. 9.4 we had a flow equal to 0.2 on the upper arc and equal to 0.8 on the lower arc, the overall cost would be equal to $0.2 \times 1 + 0.8 \times 0.8 = 0.66$.

Suppose, for example, that the overall traffic flow is equal to 1 unit (say, 1,000 cars). If everybody used the upper link in the figure, the cost would be $1 \cdot c_1(1) = 1$. Such configuration is however not an equilibrium. Indeed, if nobody used the lower link, a car driver would find it convenient to leave the first route for the second, since the cost that he would find on the new link would be equal to $c_2(0.001) = 0.001$, which is much lower than 1.

In general, the traffic is in equilibrium if for each pair of paths P and P' from s to t in the network, such that the traffic along P is positive, it holds that the cost

Fig. 9.5 The two networks of Braess' Paradox

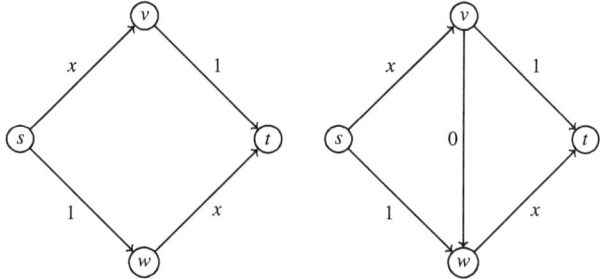

along P is not larger than the cost along P'. As a consequence, the costs of paths actually used in the equilibrium are all equal, and they are all less than or equal to the cost which a single driver would incur on any unused path.

If the traffic was equally split among the two links, the total cost would be $1/2 \times c_1(1/2) + 1/2 \times c_2(1/2) = 1/2 + 1/4 = 3/4$. But that would not be an equilibrium, since the flow along the upper link would be positive but the one along the lower link would have a cost smaller ($1/2$) than that of the upper link (1).

It can be checked that the only equilibrium in this example corresponds to the case in which the flow is only routed along the lower link. We thus obtain an overall cost equal to $1 \cdot c_2(1) = 1$. As seen above, this value is not optimal; in terms of the price of anarchy we have a ratio between cost at equilibrium and optimal cost equal to $4/3$.

The example just seen was discussed in the 1920s by the British economist Arthur C. Pigou. Afterwards it became clear that the individual behavior of the users can even give rise to some counterintuitive phenomena, as shown by the following example due to German mathematician Dietrich Braess (Fig. 9.5). Assume once again one unit of flow has to be routed. In the network on the left of the figure, the equilibrium flow is the one in which half of the traffic follows the path (s, v, t) and the other half the path (s, w, t). The overall cost is then $(1/2) \times (1/2 + 1) + (1/2) \times (1 + 1/2) = 1.5$.

In the network on the right a 'superhighway' has been added that has zero cost and connects v and w. In the new network now, the only equilibrium is the one that routes all the flow along the path (s, v, w, t). But this flow has overall cost equal to $1 \times (1 + 0 + 1) = 2$, larger than in the previous case! The apparently beneficial decision of adding a connection from v to w has thus given rise to a degradation of the system's performance. Such a phenomenon is now known as *Braess' Paradox*.

Anomalies notwithstanding, is it possible to give more encouraging results for such routing games? How large can, in general, be the price of anarchy of such games? In general, the answer depends not so much on the structure of the network, as on the type of cost functions. If such functions are linear, that is, of the form $c(x) = ax + b$, Tim Roughgarden and Éva Tardos have shown that the price of anarchy is never larger than $4/3$, so that things never get worse than in Pigou's example, even in networks that are much larger and more complex. On the other hand, if the cost functions have some "nonlinearities", then the price of anarchy can be very high and it is not possible to bound it a priori. Informally, this corresponds to saying that things tend to get worse when roads have a certain maximum capacity and more easily act as bottlenecks.

9.5 Mechanism Design and Online Auctions

Mechanism design is the field of economics that is concerned with procedures for aggregating the preferences of rational agents among a set of possible economic choices, about the allocation of goods and the definition of prices. The goal of mechanism design is to define an algorithm that takes as input the agents' evaluations and returns as output the allocation of goods to the agents and the price that the agents should pay for the allocated good. Analogously, one can view computer science as interested in algorithms and protocols that, through the use of computationally limited resources, determine solutions of good quality to a problem for a given set of input data. Mechanism design is therefore an ideal meeting ground between economics and computer science. The algorithmic design of mechanisms is a research area originated by N. Nisan and A. Ronen that has flourished in recent years and that aims at formally defining the economic features that are algorithmically implementable and the computational issues that arise in the design of economic mechanisms. To introduce some of the main ideas in the algorithmic design of mechanisms we will refer to some simple examples in the area of online auctions.

9.5.1 The Vickrey Auction

Online auctions are a typical scenario in which the design of some economic mechanism is required. The goal of a mechanism for an online auction is the identification of an allocation of the goods to the participants and the definition of the price that each participant will pay. Such an algorithm has to be conceived with the goal of satisfying some desirable features both from the point of view of the auctioneer, such as the maximization of the profit obtained by selling the goods, and from the point of view of the users, such as the perception of the electronic commerce site as reliable and of the auction as fair.

Consider first a well-known example of auction, the Vickrey auction or second-price auction, that aims at selling one indivisible good, such as an art piece, to one of a set of n interested agents. The Vickrey auction assumes that the ith agent has his own evaluation v_i for the good being sold and that he communicates to the mechanism, by means of a sealed envelope, his offer b_i for the good. Observe that user i has some private information (the evaluation), while only the offer b_i is communicated to the mechanism, that will open the envelopes after all offers has been received from all participants. The mechanism has to determine the agent i that will obtain the good and the price p_i that agent i should pay. The utility function of player j is $v_j - p_j$ in case he obtains the good at price p_j, or 0 if he does not obtain the good. We are interested in agents behaving rationally, that decide upon their strategy – the offer communicated to the mechanism – in a way that maximizes their own utility.

The strategy of an agent may in principle depend on the behavior of the other players, whose offer and valuations are however not known to the agent, and from the publicly known mechanism, deciding the allocation of the good and the selling price. What one aims to design is a mechanism for managing the auction that induces a dominant strategy for each agent, that is, a strategy that optimizes the payoff of each player independently from the behavior of other players.

An example of auction that does not induce a dominant behavior is the assignment of the good to the agent with the highest offer, for the price indicated in the offer. In such an auctions there may be several situations in which a lower offer from a player may lead to the allocation of the good for a lower price. Consider, for example, the case of two agents that offer for the good the values $b_1 = v_1 = 10€$ and $b_2 = v_2 = 20€$. Observe that in this example the offers coincide with the valuations. The strategy of the second player is not optimizing the agent's utility, since any offer above 10€ will allow to obtain the good for a lower price.

The Vickrey auction, or second-price auction, also allocates the good to the highest bidding player, but at the second highest price that is offered by any player. In the preceding example, player 2 would obtain the good at a price of 10€. One can observe that in this case, independently of the offers of the other players, the dominating strategy is to communicate an offer equal to one's own valuation for the good being sold. The selling price, and thus the utility of the agents, are not affected if a different offer from an agent does not change the allocation of the good. A different agent could obtain the good by bidding an offer higher than his own valuation, but then he would incur a negative utility. An offer lower than the valuation might cause the loss of the good and therefore a reduction in utility. Agents have therefore no incentive in communicating a bid different from their own valuation of the good.

The Vickrey auction, for all its simplicity, allows us to reason about several important properties of mechanisms:

- *Social welfare*: Firstly, the good is being assigned to the user that had the highest valuation for it. The social welfare of an allocation is the sum of the valuations of the agents for that allocation. The Vickrey auction is maximizing the social welfare, since its maximum value is equal to the maximum valuation of a single agent. Such a goal is relevant, since it is usually desirable that goods are given to the users who value them most highly.
- *Implementability*: A second important property is the possibility of each user implementing his own strategy simply, without having to ponder the other players' choices. In other words, each agent needs an elementary computational power to determine his own optimal strategy, called dominant. In the case of the Vickrey auction the dominant strategy is very simple, since it corresponds

to communicating an offer that is equal to the user's valuation. We observe that in less elementary classes of games, such as those illustrated in the preceding sections, the optimal strategies, such as pure Nash equilibria, may be multiple or not even exist, and reaching those strategies might require a set of relatively complex interactions among the agents.

- *Truthfulness*: An additional result of the Vickrey auction is the revelation from each agent of his own evaluation, in other words of the private information that defines the agent and his contribution to the input data of the problem. This property is the prevailing fundamental solution concept in the design of mechanisms, since in its absence the algorithm allocating the goods has to base its decisions on misleading information about the features of the participating agents, and thus on input data that are different from the actual ones.

9.5.2 Vickrey–Clarke–Groves Mechanisms

Some assumptions of our model of the Vickrey auction require further discussion. In particular, we assumed that the valuation of the good by an agent could be expressed in monetary terms. It follows that the utility of an agent is also expressible in monetary terms: the utility of agent i when receiving the good at price p_i is equal to $v_i - p_i$, or 0 if the good is not allocated to the agent. Such a utility model is called quasilinear. But other models have been extensively adopted during the history of economical sciences. For example, each agent could express an order of preference among all possible alternatives or communicate his own preference among each pair of allocations. However the quasilinear model of utilities yields important results in terms of dominant strategies that may not be achieved by other models.

We should also define in a slightly more formal way the notion of mechanism, in order to better appreciate the generality of this concept. The notion of mechanism in the economic sciences is the counterpart of the concept of algorithm in computer science. A mechanism for the allocation of goods has as input the description of the preferences of the agents among all possible allocations of the goods. The output of the mechanism is the selection of one allocation for the goods and of the payment that each agent should contribute. Somewhat more formally, the input to an economic mechanism is composed of a set I of n agents and a set A of possible alternative allocations of the goods. For example, in the case of an auction for a single good, the set of alternative allocations is composed of the n distinct assignments of the good to one of the n agents, the winner of the auction. The preferences of agent i on each alternative in A are represented by a valuation function v_i. The term $v_i(a)$ denotes then the valuation of agent i for alternative a. The function v_i is, however, a private information of agent i. It may be convenient for the agent i to communicate a different and not necessarily truthful valuation function v'_i. The result of the application of the mechanism, that is, the output of the mechanism, consists in the selection of an alternative in A and of a payment p_i for each agent i.

9 Algorithms for Auctions and Games

The key concept in mechanism design is that of truthfulness, as we already informally discussed in the case of the Vickrey auction. A mechanism is truthful if, for every agent i, reporting the true valuation function v_i is a dominant strategy when maximizing the utility and so is preferable to reporting any other function v'_i. If a and a' are, respectively, the alternatives selected by the mechanism when agent i declares valuations v_i and v'_i, and p_i, p'_i are the respective payments, then a mechanism is truthful if for every agent i, $v_i(a) - p_i \geq v'_i(a') - p'_i$, independently of the strategy of the other agents. Such mechanisms are also called *direct revelation* mechanisms since they are implementable by an algorithm that receives from the agents the real input data formed by the private valuation of the agents. This is not the only class of mechanisms that can implement a function as the dominant strategy equilibrium of a game. The so-called *Revelation Principle* ensures, however, that all mechanisms that can be implemented through dominant strategies can also be implemented in the form of truthful mechanisms.

The issue is then that of determining a selection mechanism for one of the alternatives and a set of payments that induce a truthful behavior from the agents. The fundamental result in this direction is the Vickrey–Clarke–Groves (VCG) mechanism, that we discuss in the related box for the interested reader.

Vickrey–Clarke–Groves Mechanism

We define the social welfare of an alternative a as the value $\sum_i v_i(a)$, that is the sum of the valuations of the agents for a. A mechanism is called a Vickrey–Clarke–Groves (VCG) mechanism (with Clarke's rule for the definition of payments) if

(i) It selects the alternative $a \in A$ maximizing the social welfare $\sum_i v_i(a)$;
(ii) It defines $p_i = \max_{b \in A} \sum_{j \neq i} v_j(b) - \sum_{j \neq i} v_j(a)$ as the payment of agent i, that is, the maximum possible reduction in social welfare of the other agents due to the existence of agent i, that by reporting valuation v_i has determined the selection of a by the mechanism.

Intuitively, the payment required from every agent is the compensation due to the maximal damage caused by agent i to the other agents because of his existence, when alternative a is selected. Such a mechanism is truthful since the agent maximizes the social welfare by revealing his own true preferences. In a way, the VCG mechanism carries out the task of decoupling the choices of the agents and determines as the dominant strategy the declaration of the true preferences of the agent. We also observe that payments are always positive and the utility of each agent equals $\max_{a \in A} \sum_i v_i(a) - \max_{b \in A} \sum_{j \neq i} v_j(b)$.

As an example of application of VCG with Clarke's rule, we proceed to prove that the Vickrey auction is a VCG mechanism. The Vickrey auction assigns the good to the agent with the highest valuation and then chooses

(continued)

(continued)

the alternative maximizing the social welfare among the n possible ones. For the agent that obtains the good, $p_i = \max_{b \in A} \sum_{j \neq i} v_j(b) - \sum_{j \neq i} v_j(a)$ is exactly equal to the second price, since the highest valuation of a different alternative is the second price offered by an agent, while the other agents have valuation 0 as they do not receive the good. For an agent i not obtaining the good, $p_i = \max_{b \in A} \sum_{j \neq i} v_j(b) - \sum_{j \neq i} v_j(a)$ has in fact value 0 since both terms equal the maximum valuation of an agent.

Another example of application of VCG is given by a multiple items auction. In a multiple items auction, k identical goods are offered to a set of $n > k$ agents, each of which seeks to obtain a single unit of the k available goods. Each agent has the same fixed valuation for all alternatives in which one of the goods is allocated to him. Each agent submits his bid in a sealed envelope. VCG chooses the alternative that maximizes the social welfare, which in this case means assigning the k units to the k agents with the highest valuations. The payments computed by VCG with Clarke's rule for the k agents that receive the good are in this case set to the $(k+1)$th highest valuation of an agent, in other words the highest valuation of an agent that does not obtain the good. Indeed, for an agent i obtaining the good, the first term in $p_i = \sum_{j \neq i} v_j(b) - \sum_{j \neq i} v_j(a)$ is exactly the sum of the $k+1$ highest valuations minus the valuation of agent i, while the second term equals the sum of the k highest valuations minus the valuation of agent i. It can be promptly checked that the payment asked to the agents that do not obtain the good equals 0. The multiple items auction can be further generalized, for example, in the setting of the so-called combinatorial auctions, that, however, we will not consider in this discussion.

9.5.3 Computational Aspects of Mechanism Design

The VCG mechanism and the definition of truthful, direct revelation mechanisms are the fundamental concepts at the basis of the area of mechanism design and its algorithmic aspects. In the following, we consider a fundamental aspect that concerns the computational issues related to social welfare maximization.

VCG requires the maximization of an objective function, the social welfare, in the domain of all possible allocations. The maximization of such an objective function can be computationally difficult for many problems of high relevance (see also Chap. 3). It is customary in the field of optimization algorithms to circumvent the computational complexity of exactly optimizing a function by the use of an approximation algorithm, that is, an algorithm that always allows one to obtain, on any instance, a solution close to the optimal one, while incurring a polynomial computational cost. In a strategic setting it is also required, however, that such functions be implementable through some dominant strategy equilibrium.

A fundamental question is thus the characterization of the social choice functions that are implementable. Are there simple properties that a function should have so that it is implementable by dominant strategies of the agents?

A relevant case is that in which the valuation of an agent is some numerical value in a subset of the alternatives $W_i \subseteq A$, called winning for agent i, while it is zero on all other (losing) alternatives. More precisely, the agent is characterized by a valuation t for all alternatives in W_i and 0 for all alternatives outside W_i. In this particular case it is possible to completely characterize the set of truthful mechanisms by relying on the monotonicity properties of the implementable functions. A single-parameter allocation function is called *monotone* in v_i if it is possible to define a single critical value corresponding to the minimum valuation that allows the agent to be in a winning alternative. For example, in the case of the Vickrey auction the critical value equals the largest of the valuations of the losing agents.

The definition of critical value plays a fundamental role in the definition of payments, since the agents' payments can be fixed exactly at the critical value. The importance of monotone functions is in the fact that a monotone social choice function and the respective critical values allow one to implement a truthful mechanism in which the winning agents pay exactly the critical value. Such a characterization is of great importance, since it allows the implementation of truthful mechanisms for the optimization of social welfare functions that are computationally hard and that can be approximated by monotone functions that are simple to compute.

Consider, for example, a multi-unit auction problem, in which m units of the same good are available to n agents. Each agent j desires some number q_j of units, known to the mechanism, and has a private valuation function v_j for obtaining such a set of units. This is a clear case in which agents are completely described by a single private value, their valuation. The problem of maximizing social welfare in the allocation of the m available units to the n agents corresponds to the well-known knapsack problem (see Sects. 2.3.2 and 5.3), for which an application of VCG would require a computational cost that is not polynomial in the size of the input. For the interested reader, the box "Monotone Mechanisms" reports an alternative approach to the same problem.

Monotone Mechanisms

We consider an alternative approach to the multi-unit auction problem, based on the adoption of monotone mechanisms that yield a good approximation to the optimal solution of the problem in question. In particular, consider these two algorithms:

1. VAL: Sort the agents non-increasingly by their valuation v_j;

(continued)

(continued)

2. DENS: Sort the agents non-increasingly by their density (valuation per unit) v_j/q_j.

Both algorithms define a ranking among the agents and allocate the elements according to this ranking, until some agent requires more units than those that remain available. Let's look at the execution of both algorithms when there are 3 players ($n = 3$), 4 units of the good ($m = 4$), and the agents' data are the following:

$$v_1 = 5, q_1 = 2; v_2 = 3, q_2 = 1; v_3 = 4, q_3 = 2.$$

The algorithm that sorts by valuation will allocate 2 units to agent 1 and 2 units to agent 3, with an overall social welfare of 9. The algorithm that sorts by density will allocate 1 unit to agent 2 and 2 units to agent 1 for an overall social welfare of 8. Observe that in this last case one unit of the good is not allocated.

Firstly, we observe that both algorithms are monotone. Indeed, if an agent with a given valuation is selected, he is also selected when his valuation is increased.

Let us compute the payments for the agents in both cases. Such payment for each agent is equal to the minimum valuation that would allow the agent to be selected. In the case of ranking by valuation, the payments required from the two winning agents equal the valuation of the losing agent.

In the case of ranking by density, consider the first agent in the ranking that is not winning. Let this agent be j. The payment of winning agent i will be equal to $q_i(v_j/q_j)$. Observe that in case all agents are winning each payment equals 0, since the presence of agent i is not decreasing the social welfare of the remaining agents.

Each algorithm individually is not producing a good approximation of the optimum. Indeed, the ranking by valuation could accept an agent requiring all m units with a valuation of 2 and at the same time not accept m agents each requiring one unit with valuation 1. Analogously, the ranking by density could accept an agent requiring 1 unit with valuation 2 and not accept an agent requiring m units with valuation m. Notice, however, that if we define a third algorithm called MAX that outputs the best solution among those given by VAL and DENS, then we obtain a solution close to optimum. In our example, MAX returns the solution constructed by algorithm VAL. It can be easily shown that, in fact, algorithm MAX always constructs a solution with value at least half the optimum solution.

(continued)

> **(continued)**
>
> Unfortunately, the algorithm obtained by the combination of two monotone algorithms is not necessarily monotone. For this to hold, it is required that the two algorithms satisfy a condition stronger than monotonicity, called *bitonicity*. The combination of two bitonic algorithms is indeed monotone. An algorithm is bitonic if it is monotone and whenever a losing agent increases his valuation then one of the two following conditions is satisfied:
>
> 1. The agent becomes winning; or
> 2. The agent remains losing but the value of the constructed solution does not improve.
>
> In the case in question, it is immediate to verify that any algorithm that defines the winning agents by traversing a ranking in which the priority of each agent can only increase if his valuation increases, is bitonic. Indeed, if the valuation has not increased enough to make the agent winning, then the constructed solution does not change.

9.6 Price-Setting Mechanisms and Competitive Equilibria

Mechanisms for defining prices of goods have played a central role in economic theory during the last two centuries. Price-setting mechanisms have been given the task of introducing fundamental notions such as efficiency and equilibrium in markets, the availability of goods and services, and the operation of the economy in stable conditions. The role of initiator of the mathematical theory of markets is attributed to Léon Walras, who in 1874 was the first to define the notion of *competitive equilibrium*: an equilibrium in which each agent obtains the good from which he extracts the highest utility and in which all goods left unsold have zero price. Mechanisms for price setting form a further fascinating field of application of algorithmic methods and provide important computational questions.

Consider the case in which we have a set I of n agents and a set J of m distinct goods. Each agent has a valuation function v_i for each set S formed by some of the m goods. The goal is to allocate a set of goods to each agent. A price-definition scheme defines a price p_j for each good $j \in J$. The *demand of agent i* is defined as the set S of goods preferred by agent i, the one that maximizes the utility $v_i(S) - \sum_{j \in S} p_j$ of the agent.

A *Walras equilibrium* is defined as a set of prices in which each agent receives his demand and all goods not allocated have price 0.

A first important result, also called *first welfare theorem*, proves that a Walras equilibrium determines an allocation maximizing the social welfare of the agents. In the following, we consider an example scenario.

Consider two agents, Alice and Bob, and two elements $\{a, b\}$. Alice has valuation 2€ on every nonempty set of elements, while Bob has valuation 3€ on the whole set $\{a, b\}$ and valuation 0€ for each single item. The solution that maximizes social welfare assigns the set $\{a, b\}$ to Bob. Thus, to obtain a Walras equilibrium, Alice should prefer the empty allocation to each of the individual items. For this to be true, the price of each good should be at least 2€. But then the price of the set $\{a, b\}$ is 4€ and so Bob, too, will prefer the empty set as an allocation. There is therefore no Walras equilibrium in this case.

An important question in economic theory is the characterization of markets that admit a Walras equilibrium. The answer to such a question can be reduced, surprisingly, to the solution of some important algorithmic questions. It is in fact possible to relate the existence of a Walras equilibrium to the existence of an integer solution to a linear optimization problem in fractional variables. We stop here and do not embark on a field that would require a long discussion. The interested reader will find in Sect. 9.7 some references to the main monographs surveying the area.

9.7 Bibliographic Notes

The book that marked the birth of game theory is the classic text of von Neumann and Morgenstern [108]; it is remarkable that the book is largely due to John von Neumann, one of the fathers of the digital computer era, as a well as world-class mathematician.

Modern treatments of the theory of games and its applications to economics are given, for example, by Binmore [9], Osborne and Rubinstein [88] and Mas-Colell, Whinston and Green [76]. The encyclopedic work by Aumann and Hart [6] is an updated state of the art of the theory, and contains many further references to the scientific literature. Computational aspects and applications of game theory to computer science are discussed in a recent textbook by Nisan et al. [86], to which we point the interested reader for further investigation of many of the topics touched in this chapter, including algorithmic mechanism design.

Chapter 10
Randomness and Complexity

Riccardo Silvestri

Abstract Random choices have an unexpected power. From a database search to traffic analysis of the Web, from data mining to cryptography, several hard problems can be efficiently solved with the help of probabilistic algorithms. But random choices are also very elusive. If they were too powerful, some cryptographic algorithms used daily would no longer be trusted. At the heart of this phenomenon, the interplay between randomness and complexity creates a fascinating world that is almost entirely unexplored. This is a world where the most sophisticated algorithms meet their analytical limitations, and the reasons for their effectiveness in real applications still remains a mystery.

10.1 A Dialogue

Francis I have a tricky computer problem that comes from a commissioned job.
Laura, Mark Tell us about that!
Francis So, my company must implement an information system on the annual consumption in Italy. The raw data to be processed are collected in a large number of files, each of which contains all commodity items that were sold in 1 year by a certain set of retail stores.
Laura Wow! It'll be a huge amount of data.
Francis Definitely. Assume that each file collects data from about a 1,000 stores, and each store, on average, sells around a 100,000 items a year (data collected refer only to medium–large-sized stores). Thus, each file contains about

---

R. Silvestri (✉)
Dipartimento di Informatica, Sapienza Università di Roma, via Salaria, 113, 00185 Roma, Italy
e-mail: silvestri@di.uniroma1.it

100 million records,[1] one for each item sold, and takes several gigabytes[2] on disk. Now, before uploading the data into a DBMS[3] (which will require a lot of time), we'd like to quickly get information that will help us to design the database.

Mark   What kind of information?

Francis   For example, we could wish to know whether a given archive contains data relating to at least a food store. I thought about it, but I didn't find anything better than the trivial algorithm that scans all the records in the archive. This requires at least a minute for scanning an archive and then it'll take a few hours to search through all the archives, since there are a few hundred.

Mark   Are items in the archives arranged in some way? For example, by store? If so then you could...

Francis   No, I thought of that too. Unfortunately, in an archive the items, namely the records, are collected as they are sent by the stores. Some make monthly transmissions, others every 6 months, still others yearly. I don't think there's any order that you can exploit to speed up the search.

Laura   Maybe you could use a probabilistic approach.

Francis   What do you mean?

Laura   Well, if you wish to get an estimate of the percentage of spoiled nuts in a bag, you don't need to open them all. You can pick at random a small sample and open just those nuts.

Mark   That's true! I think I understand: the nuts are the items, the bag of nuts is an archive and the spoiled nuts are the food items.

Laura   Exactly.

Francis   What? What are you talking about!?

Laura   You're right, Fran, let's talk about details. We wish to know whether in a given archive there are items sold by food stores. You said that the average number of items per store is about a 100,000. I think it's likely that the food stores are among those which sell a large number of items, and so the number of items sold from a food store is not lower than the average. An educated guess is that any food store in the archive has sold at least a 100,000 items.

Francis   Yes, I agree with your guess. But so what?

Laura   All right. If the archive contains at least one food store, and I pick at random an item in the archive, what's the chance that the picked item is a food item?

Francis   Ahem. It should be easy. There are about 100 million items in the archive, and among them, food items number at least a 100,000. So there are at least 100,000 favorable cases out of 100 million. The probability is at least

---

[1] A record is an elementary data item in a database.

[2] A gigabyte is a unit of measure of the volume of data and is equivalent to approximately one billion bytes.

[3] A DBMS (Data Base Management System) is a software system that manages archives or databases.

$$\frac{100{,}000}{1{,}000{,}000{,}000} = \frac{1}{1{,}000}.$$

A chance out of a thousand, I'd say rather small.

Laura But not too small. If we choose, again at random, k items instead of just one, what's the probability that at least one of the chosen items is a food item?

Francis Why is it always up to me? Okay, I'll put my best foot forward. So, in such cases, I think, it's easier to calculate the opposite probability, that is, the probability that none of the chosen items is a food item. For any chosen item, the probability that it is not a food item is at most

$$1 - \frac{1}{1{,}000}.$$

Since the choices are independent,[4] the probability that none of the k choices succeeds in finding a food item is equal to the product of the individual probabilities, and then it's at most $(1 - 1/1{,}000)^k$. At last I got it: the probability that at least one of the chosen items is a food item is at least $1 - (1 - 1/1{,}000)^k$.

Laura Now simply impose that k be a sufficiently large value (but not too large) so that the probability is very high. You could set $k = 30{,}000$ to obtain

$$1 - \left(1 - \frac{1}{1{,}000}\right)^{30{,}000} \geq 1 - 10^{-13}.$$

So, the probability that by examining 30,000 randomly chosen items you don't get the correct answer is less than

$$10^{-13},$$

that is, one in 10,000 billions.

Francis Now I understood the algorithm: we examine 30,000 records (items) chosen at random; if at least one is a food item we conclude that the archive contains a food store, otherwise we bet that it doesn't. The probability that the algorithm gives a wrong result is very small. In fact, it is a thousandth of the probability of winning the lottery!

Mark And that probabilistic algorithm is much faster than the trivial algorithm that looks at all records. Indeed, it only needs to read 30,000 records rather than 100,000,000. I guess it's at least 3,000 times faster.

---

[4] In probability theory the concept of stochastic independence plays an important role (see box "Events, Probabilities and Expected Values" in Chap. 9). Here we can say, informally, that the independence of choices means that there are no relationships among the choices: each of them is carried out in a completely independent way from the others.

Francis The estimate of how much it's faster than the trivial algorithm is not so easy, because the time of reading records scattered here and there in a file is well above the time to read contiguous records. Anyway, I'd expect that the probabilistic algorithm is 100 times faster. This means that instead of waiting hours to get the answer, I'll get it in a few tens of seconds. That is really a great idea, Laura!

Laura Thanks, but the idea is certainly not new. In one way or another it's used in a lot of algorithms. The general technique is called *random search*, and it's used to check polynomial identities and in several algorithms of number theory, some of which have applications in cryptography.[5]

Francis I didn't know it. It's very interesting. I'm wondering whether similar techniques can help me to solve other problems related to my archives. One of those is that we'd like to estimate the number of distinct types of items. Each record contains a string that describes the type of the item. For example, "hammer", "bread", "electric toothbrush", etc. We'd like to know, approximately, how many different types appear in the archives. This kind of information would be very useful for improving the design of the database. In addition, getting this information will not be easy, even after we have the database. I mean, it still might require a fairly long computing time.

Mark How long can that be?

Francis The algorithm I thought of is based on building the sorted list of all the item's types, and then scrolling through the list to determine the number of different types. But, given the large amount of data, sorting takes a long time because it needs to be done mostly on the disk. I estimate the algorithm could take several dozens of hours.

Mark Days of computations.

Francis Oh, yeah. Also, keeping the sorted list of strings requires a great deal of memory on disk.

Laura Your problem, Fran, can be solved by *probabilistic counting*. It's an algorithmic technique that was specifically invented to solve problems like yours. In fact, it's used in data mining,[6] for optimizing queries in databases, and to monitor the traffic in communication networks.

Francis Is it a difficult technique, or you can explain it easily?

Laura The basic idea is pretty simple. In its mathematical essence, the problem is to determine, or, better, to estimate the cardinality of a multiset.[7] Let's first consider the following simplification. Suppose that the elements of our multiset M are binary strings of length 30, selected at random. I'm assuming that the

---

[5] See Chap. 6.

[6] *Data mining* is a discipline of computer science that deals with the exploration, extraction and analysis of information implicitly contained in large databases.

[7] The term multiset means a collection of elements that may contain repetitions of the same element. In our setting, by cardinality of a multiset we mean the number of distinct elements belonging to the multiset.

types of items are described by binary strings of length 30 and that these have been chosen at random among all the binary strings of length 30. But I make no assumptions about their repetitions: each string can be repeated, in the archive, an arbitrary number of times. Now, in order to estimate the cardinality of M, we can exploit the fact that the distinct elements of M are randomly distributed.

Francis Oh, is that true? And how can you?

Laura Suppose you scan all the strings of the multiset M and check whether some fall in the set P_{10} of strings that have the first symbol 1 in position 10. What's the probability that no string falls in P_{10}?

Francis If I understood correctly, I should calculate the probability that a random set of strings doesn't contain strings beginning with 0000000001. The set P_{10} contains exactly

$$2^{30-10} = 2^{20}$$

strings. So the probability that a string chosen at random, among those of length 30, doesn't fall into P_{10} is

$$1 - \frac{2^{20}}{2^{30}} = 1 - \frac{1}{2^{10}}.$$

Letting n be the number of strings, the probability that none of the n strings falls into P_{10} is equal to the product of the probabilities, that is

$$\left(1 - \frac{1}{2^{10}}\right)^n.$$

Mark No! My dear friend, you implicitly and incorrectly assumed that the choices of the n strings are independent. That's not true because there's the constraint that they're all distinct. Anyway, if we assume that n is much smaller than 2^{30}, the probability that you've calculated is a good approximation to the correct one, which would require a fairly complicated expression.

Francis What a dunce I am!

Mark Don't worry, you're in good company. The probability of making mistakes calculating probabilities is very high.

Laura Let's not digress. Then, using the probability that Fran has calculated, we note that, for $n \leq 200$, the probability that P_{10} is empty is greater than

$$\left(1 - \frac{1}{2^{10}}\right)^{200} > \frac{4}{5},$$

while for $n \geq 2{,}000$ that probability is less than

$$\left(1 - \frac{1}{2^{10}}\right)^{2{,}000} < \frac{1}{7}.$$

So if we see that, after scanning all the strings of the multiset, P_{10} remains empty we bet that $n < 2{,}000$, while if we see that it's not empty then we bet that $n > 200$.

Francis I'm beginning to understand. By keeping just one bit of information, that is, whether P_{10} is empty or not, we get information on how large n is.

Laura Exactly. Let's consider now all the brother sets of P_{10}: P_k is the set of strings of length 30 that have the first 1 in position k, for $k = 1, 2, \ldots, 30$. These sets are a partition of the set of binary strings of length 30, except for the all-0 string. Moreover, each of them is twice the size of the next one: P_1 is twice the size of P_2, which is twice the size of P_3 and so on. We keep a bit b_k for each P_k that, after scanning, will be 1 if P_k is not empty, and 0 otherwise.

Francis So at first, all those bits are set equal to 0 ($b_1 = 0, b_2 = 0, \ldots, b_{30} = 0$) and then every time a string is scanned, if it has the first 1 at position k, let $b_k = 1$.

Laura Yes. Let's call R the maximum k for which $b_1 = 1, b_2 = 1, \ldots, b_k = 1$, after the scan is over. It can be shown that R is an approximation of $\log_2(n)$ and thus 2^R is an approximation of n. Clearly, this approximation is quite rough.

Francis If I understand correctly, by keeping just 30 bits of information while scanning the elements of the multiset, you can get an approximation of the cardinality, although a little rough. Remarkable!

Laura Not only that. What I have told you is the basic version of the technique. A refinement leads to an approximation with an error of less than 3%, while keeping the auxiliary memory used for the counting as small as a few thousands of bytes. And that even if the cardinality is of the order of billions.

Francis Wonderful! This technique would allow me to get a good estimate of the number of types of items in just a few hours, which is the time to perform a full scan of all archives. Also, if I want an estimate relative to some other characteristic (for example, brands), I can do that at the same time, because the memory required for each count is very small. With a single scan I can do a lot of counts at the same time, at the price of only one!

Laura Exactly! There are instances in which the use of this algorithmic technique led to a reduction in the computation time by a factor of 400. Specifically, it was an analysis of the Web graph, that is, the connections by links among the pages of the Web.[8]

Mark Excuse me, but aren't you rushing a bit too much? Laura, isn't something missing in the description of the technique? Everything that was said is valid under the assumption that the set of distinct elements have been chosen uniformly at random. Or not?

Laura Sorry, in the rush to tell you the results of the technique I forgot that point, which is indeed very important. Clearly, in practice that hypothesis is not likely. Especially considering that the domain from which the elements are drawn has

---

[8] See Chap. 5.

to be sufficiently simple that it can be subdivided in an efficient manner, as the set of binary strings of fixed length.

Francis   Ah! It was too good to be true.

Laura   Don't worry, Fran. It's enough to use a hash function[9] H which assigns to each element of the multiset (whatever its nature) a binary string of suitable length L. An ideal hash function assigns to each element a randomly chosen string of length L. So, for every element x, $H(x)$ will be a random binary string assigned to x. A way to describe such a hash function is as follows. Whenever you need to compute $H(x)$, check if $H(x)$ has already been computed, if so return that value, otherwise pick at random a binary string of length L and this will be the value of $H(x)$. By using a hash function we can turn any set into a random set of binary strings. And the probabilistic-counting technique, which we have discussed, applies to the multiset of binary strings obtained through the hash function.

Mark   And the collisions? What happens when two distinct elements x and y are turned into the same binary string $H(x) = H(y)$?

Laura   If you choose the length L of the strings large enough, the collisions will have a negligible weight on the probabilistic counting. For example, if you want to estimate a cardinality up to a few billions, simply set $L = 64$ to make the chance of collision so small as to be infinitesimal.

Mark   The problem still remains of calculating the ideal hash function that you described. You know perfectly well that in order to implement that function you should keep a table of size proportional to the cardinality that you want to estimate, not to mention the time to look up values in this table. So, it would defeat all or most of the advantages of the technique.

Laura   Yes. That's why in practice nonideal hash functions are used. They're easy to compute and work great. Using those the technique retains all the advantages that we have seen, both in terms of computing time and storage space.

Francis   I'm really confused right now! Although, as you know, I'm more focused on practice than theory, I thought I had understood that the assumption on the randomness of the set is needed for the technique to work. So much so that it's necessary to use an ideal hash function to turn any set into a random set. But now Laura says you can use a nonideal hash function which is computable by an efficient algorithm. How does this function guarantee that the set is turned into a random set?

Laura   In fact, it doesn't guarantee it. Besides, what does it mean to say that a set is random? In this regard, I'd recall what Knuth wrote

> It is theoretically impossible to define a hash function that creates random data from non-random data in actual files. But in practice it is not difficult to produce a pretty good imitation of random data.[10]

---

[9]Hash functions are widely used in computer science with applications in databases, encryption and error correction.

[10]See [69].

Of course, that doesn't satisfactorily answer your question. I think actually there's a lack of analysis of the technique. But...

Mark   Excuse me for interrupting you, but the question raised by Francis is extremely delicate, fascinating and much broader than you might imagine. I fear that if we insist on staying on the question we run the risk of falling into a well of unknown depth. Maybe we can come back to it later. But now I've come up with an idea on probabilistic algorithms I wish to discuss with you. Is it true that probabilistic algorithms, which can go wrong, albeit with an extremely low probability, are only used in situations in which a possible error would not be too harmful?

Laura   You think so? Actually it isn't. When the error probability is low enough, with respect to the cost of a possible error, and the speed of the algorithm is a very important aspect, then the probabilistic algorithm is used, even if the cost of a possible error is very high.

Mark   Really?

Laura   Definitely. Every day, lots of financial and commercial transactions take place on the Internet, and these are secured by communication protocols that encrypt transmitted messages. The most widely used encryption protocol uses the asymmetric encryption algorithm RSA.[11] The RSA algorithm relies on the choice of very large prime numbers, with hundreds of decimal digits. To do that, an integer of the required size is generated at random and then it's checked for primality. The procedure is repeated until a prime number is found. The primality tests currently used are probabilistic algorithms.

Mark   But I know that, a few years ago, a deterministic primality test was invented that's not probabilistic and thus is always correct. If I remember correctly, it's also efficient.

Laura   Yes, but it's not efficient enough. You can't afford to wait several minutes for a transaction to be carried out, especially when there are tens of thousands of transactions per day. Think about an online bank.

Francis   I'm not an expert like you on probabilistic algorithms and I'm curious to find out more about probabilistic primality tests.

Laura   One of the most widely used is the Miller–Rabin primality test.[12] To explain the idea of the test, I'll start from the simplest primality test: given an integer n, for every integer $x < n$ ($x > 1$) check whether x divides n, if it does then n is not prime; if no x divides n, then n is prime. When we find an x that divides n we say that x is a witness of the non-primality of n, or compositeness witness for n. If n is prime, there are no compositeness witnesses, while if n is composite, there is at least one.

Francis   For instance, if $n = 15$, then 5 is a compositeness witness for 15 and also 3 is, but 2 is not. Instead, if $n = 13$, no compositeness witness for 13 exists because none of the integers $2, 3, \ldots, 12$ divides 13. Right?

---

[11]See Sect. 6.6.

[12]The test was invented by Gary L. Miller and Michael O. Rabin.

Laura   Exactly. However, this test is too inefficient even if we can improve it a lot by noting that we can limit the search for the divisors of n among the integers not greater than the square root of n. If n has 100 decimal digits the test can require about 10^{50} divisions. Putting together all the computers of the planet, a 1,000 years would not suffice to complete the calculation.[13]

Francis   Oh boy!

Laura   Yeah. And that's where the power of the Miller–Rabin algorithm helps. Rather than using compositeness witnesses based upon divisibility, the Miller–Rabin test uses a much more refined kind of compositeness witnesses. I won't go into details about how these compositeness witnesses are defined because it could distract us from the probabilistic structure of the test. Suffice it to say that, for every n and for every x, we define a property $MR(n, x)$ and if it's true then x is a compositeness witness for n. Indeed, it was proved that if n is prime then no x makes $MR(n, x)$ true, and if n is composite at least an integer x exists which makes $MR(n, x)$ true.

Francis   If I understood correctly, that property can be used in place of the one based on the divisibility. But what's the advantage?

Laura   The advantage is that when n is composite not only can we say there is at least one witness, but that there are a lot of them. To be precise, at least $3/4$ of all possible x are compositeness witnesses, that is, they make $MR(n, x)$ true. So if you pick an x at random in the interval $[1, n-1]$, it'll be a compositeness witness for n with at least a $3/4$ probability. In other words, with one try the error probability is at most $1/4$. By k retries the error probability decreases to

$$\left(\frac{1}{4}\right)^k.$$

Usually k is set equal to 50, so the error probability will be less than

$$2^{-100}.$$

That probability is so small that it is easier to win the lottery three times in a row rather than to fail the Miller–Rabin test.

Mark   Oh yes, it's clear. Moreover, that error probability is so small that it is comparable to the probability of the occurrence of a hardware fault during the execution of the algorithm. Then, probabilities of error so small make deterministic algorithms indistinguishable from the probabilistic ones, as far as their ability to provide correct answers.

Francis   So, the Miller–Rabin test still uses the random search technique, as in the case of my problem, and the error probability is negligible. But, how much faster is it than the simple test based on divisions? Making an analogy with what

---

[13] See Sect. 6.6.

we saw in relation to my problem, I'd say a few hundred times, maybe some thousands?

Laura Are you kidding? For numbers with 100 decimal digits, the Miller–Rabin test is about

$$10^{45}$$

times faster than the test based on divisions!

Francis 10^{45}?! I can't even remotely imagine a similar speedup.

Mark You're right ... who can?

Francis But, a terrible doubt entered my mind just now: How does a computer make random choices?!

Mark Your doubt is perfectly legitimate. It's enough to remember what von Neumann said in this regard, more than half a century ago:

> Anyone who attempts to generate random numbers by deterministic means is, of course, living in a state of sin.[14]

A mathematically rigorous meaning to this statement can be given by the Kolmogorov complexity theory.[15] Leaving out many details (which indeed are not just details), I could explain the idea upon which the theory is based, in a nutshell. Suppose you toss a coin 1,000 times and record the sequence of heads and tails. What is expected, and indeed it can be proved, is that with very high probability the sequence has no meaningful regularities. We don't expect to find that, for example, every three tosses there's at least one head, or that there's a subsequence of 50 consecutive tails, or that the number of heads is substantially higher than the number of tails, etc.

Laura But how can the concept of regularity be defined in a formal way? It should include all kinds of regularity and I don't see any regularity shared by all the regularities.

Mark That's right! It would be very hard, if not impossible. Kolmogorov didn't directly use the regularities but rather a consequence of their presence. If the sequence has any significant regularity, then it can be exploited to give a compact description of the sequence, which is more compact than the sequence itself. The description can be given by an algorithm (in the theory, the descriptions are precisely algorithms) whose output is the sequence itself.

Laura I think I understand. Suppose, for instance, I've a sequence of 10,000 bits such that all the bits in the even positions have value 1 and the others have random values. Then I can describe the sequence using a simple algorithm that always outputs a 1 if the bit is in an even position, and otherwise it outputs the bit that it reads from a table containing only the bits in the odd positions of the sequence. The algorithm has a description whose length is only slightly greater than half

---

[14]See [69].
[15]See Sect. 7.3.2.

the length of the sequence, and so it's much more compact than the description given by the sequence itself.

Mark That's right. Now, it's not hard to prove that a sequence of random bits, with very high probability, does not have a substantially more compact description than that of the sequence itself. Summing up, we can say that random sequences do not have compact descriptions. So, if a sequence has a compact description, it's not a random sequence.[16]

Francis Okay, beautiful theory. But, what does it have to do with the possibility that computers could make random choices or not?

Mark It's very simple. If someone says that he found an algorithm that's able to generate random sequences, then it's easy to refute it. Make the algorithm generate a sequence substantially longer than the description of the algorithm. This sequence has a description that's obtained by combining the description of the algorithm with the description of its length. It's more compact than the sequence itself. Thus, the sequence cannot be considered to be random. More precisely, it can't be considered as if it were generated by a genuine random source.

Francis Gosh! I understood. But then there's no hope.

Laura Not really, there are algorithms that are able to extract from the computer (for example, by reading the microseconds from power on, the current number of reads from the disk, the number of currently active processes, etc.) a small amount of genuinely random bits and then, through appropriate processing, they can amplify them, producing a much longer sequence of random bits.

Mark But also in that case, Laura, we can apply Kolmogorov's theory. It's enough to consider in the description the random bits derived from the instantaneous state of the computer, in addition to the algorithm and the length of the sequence. The truly random bits that can be drawn are few compared to those required and then the amplification cannot be too small, so the description will be compact.

Laura Oh! It's true. Indeed that type of algorithm, called a pseudorandom generator, was developed for cryptographic applications. And the properties that they must meet are captured by rigorous mathematical definitions. The situation is quite different from that of probabilistic algorithms. Yet there are strong connections, but it would be a long story. Instead, I'd like to point out that the implementations of probabilistic algorithms often use very simple generators. For example, among the most simple and the most widely used generators there are the so-called *linear congruential generators* that have the form:

$$x_{i+1} = a \cdot x_i + c \pmod{m}$$

---

[16] The term *random sequence* is used here in an informal way, hoping that it will not be too ambiguous or, worse, misleading. It is clear that no sequence can be said to be random or not random in the sense that all sequences, of a fixed length, have the same probability to be generated by a genuine random source (uniform), such as repeated coin tosses.

where a, c and m are integer parameters. The sequence of pseudorandom numbers is started by the set value x_0, called seed. Then, the successive numbers are computed by applying the formula to the previous number. A possible set of parameters is as follows: $a = 16,807$, $c = 0$ and $m = 2,147,483,647$. Despite their simplicity, I'm not aware of discrepancies that have been observed with respect to what would be expected if genuine random sources were used instead of such generators.

Mark I don't want to be a spoilsport, but I know at least one case in which such discrepancies were observed. In 1992, a computer simulation of a simple mathematical model of the behavior of atoms of a magnetic crystal didn't give the expected results. The authors of the simulation showed that this discrepancy was just due to the pseudorandom generator that was used. They also noticed that many other generators, among the most used and which passed batteries of statistical tests, were affected by similar flaws. One can say that those generators are poor imitators of truly random generators. However, we should also keep in mind that this incident concerned a simulation. I don't know similar incidents concerning probabilistic algorithms.

Francis This heartens me. If I understand your conversation, I could summarize the situation (paraphrasing a famous quote by Eugene Wigner[17]) talking about the unreasonable effectiveness of *deterministic* pseudorandom generators to imitate truly random generators.

Laura To sum up, in theory and especially in the practice, probabilistic algorithms work great. And then I wonder what's, in general, the power of probabilistic algorithms? What's the power of "random choices"? Maybe, for every hard problem there's a probabilistic algorithm that solves it much faster than any deterministic algorithm.

Mark If so, we could be in trouble.

Francis But how!? We might be able to solve lots of problems that now we don't know how to solve.

Mark Of course, but it also would happen that the most widely used algorithms and protocols to secure Internet communications would be completely insecure. In addition...

Francis Wait a moment, but there's a mathematical proof of security for those algorithms, isn't there?

Mark No, currently there's no absolute guarantee of their security. Their claimed security relies on a tangled skein of empirical data, assumptions and conjectures.

Francis Holy cow! Should I be more careful when I use my credit card on the Web?

---

[17]Eugene Paul Wigner was one of the greatest physicists and mathematicians of the last century (he won the Nobel Prize for Physics in 1963). The phrase is actually the title of his famous essay: *The Unreasonable Effectiveness of Mathematics in the Natural Sciences*.

Mark   If someone knew a way to defeat the protection provided by the present cryptographic protocols, I don't think he would waste time with your credit card. He would have at his disposal much richer targets before being found out.

Laura   That's really true. An example of such protocols is once again RSA.

Mark   Yeah, and the interesting thing is that the security of RSA relies on a problem that seemingly is very similar to the problem solved by the primality test.

Francis   Ah! Tell me.

Mark   The security of RSA relies on the (conjectured) difficulty of the integer factoring problem. Given an integer n, the problem is finding all the prime factors.[18] Actually, we can consider an apparently simpler version: given a composite integer n, find a nontrivial divisor (that is, different from 1 and n) of n. If you know how to efficiently solve this version of the problem, you also know how to efficiently solve the full version.

Francis   Well, the algorithm based on the divisions that we saw for the primality test also solves this problem. When n is composite, it can be stopped as soon as it finds the first nontrivial divisor.

Mark   Yes, of course, but it's not efficient. And it's totally inefficient when n doesn't have small divisors, and this happens, for example, when n is the square of a prime number.[19] Not by chance, the security of RSA relies on the (conjectured) difficulty of factoring a number n which is the product of two large prime numbers (that is, with the same number of digits). On the other hand, the Miller–Rabin algorithm, which is so efficient to test primality, when applied to a composite number, does not provide significant information about possible divisors of n.

Laura   Yes, the compositeness witnesses of Miller–Rabin are very different from those of the test based on divisions. The latter directly provide a divisor of n, while those of Miller–Rabin are indirect witnesses: They ensure that at least a nontrivial divisor exists but don't provide meaningful information about it. On closer look, right here is the power of the Miller–Rabin algorithm.

Francis   It's strange: You can guarantee that a thing exists without exhibiting it and without even stating an easy way to find it.

Mark   You don't know how much you're justified in saying that it's strange. That strangeness originated a long time ago; think about, for instance, the controversy about constructivism in mathematics at the beginning of the last century.[20] Or, closer to our interests, consider the so-called probabilistic method that is a technique of proof which draws its strength from the opportunity to prove that a thing exists through a probabilistic argument that doesn't exhibit the thing itself.

Laura   I'm sorry to interrupt you now, but I wouldn't want to get lost, as you said, in a bottomless well.

---

[18] A prime factor is a divisor that is also a prime number.

[19] Obviously, in this case, it would be very easy to factorize n: just compute the square root of n.

[20] See Chaps. 1 and 3.

Mark Actually I talked about a well of unknown depth; I think there's a subtle difference. Anyway you're right, back to the factoring problem. The present situation can be summarized by saying that over recent decades, thanks to the introduction of RSA and its growing importance, various techniques and algorithms for integer factoring have been developed and then steadily improved. These algorithms are much more efficient than the algorithm based on divisions, but they're still inefficient. I mean that they can't factorize integers of hundreds or thousands of digits within a reasonable time. The best algorithms (which indeed also require human intervention in setting up some critical parameters based on preprocessing of the integer) have recently factorized an RSA-type integer of 200 decimal digits within 5 months of calculation using several computers simultaneously. This is a result that maybe a dozen years ago, would not have been foreseeable.

Francis I get goose bumps. But then where does the confidence in RSA come from? I've heard about NP-completeness,[21] maybe it has something to do with this?

Mark Yes and no. NP-complete problems are considered difficult to solve because it's believed that the conjecture NP $\neq$ P is true. The factoring problem is not NP-complete, or, better, it's not known whether it's NP-complete or not. However, if the conjecture were not true, then there would be an "efficient" algorithm for factoring. I say "efficient" in quotes because the fact that the conjecture is false doesn't imply that such algorithms are necessarily efficient in practice. I don't want to go into this issue because the discussion would be much too long. However, even if the conjecture were true and the factoring problem were shown to be NP-complete, this doesn't necessarily guarantee the security of RSA.

Laura I don't understand. If NP $\neq$ P and the factoring problem were NP-complete, then it would be guaranteed that efficient algorithms for factoring cannot exist.

Mark Yes that's true, but the theory at issue says nothing about the possibility that there could be algorithms that are efficient on a subset of instances only. I'll explain: even if what we have supposed were true, there could be an algorithm that is able to efficiently factor a substantial fraction of all integers in the sense that this possibility is perfectly compatible with the present theory. And this would be more than sufficient to make RSA totally insecure.

Laura You're right, in order for RSA to be insecure it is sufficient that there is an algorithm that can efficiently factorize a small fraction of the integers of the type used in RSA. For the overwhelming majority of the numbers it could be totally inefficient. In addition, the algorithm could be probabilistic.

Francis I don't have your knowledge on this topic and I can only figure out that the situation is like a tangled web. I'd like you to better explain the phenomenon,

---

[21]The theory of NP-completeness is discussed in Chap. 3.

10 Randomness and Complexity

rather surprising to me, that there are difficult problems that can be efficiently solved on many instances.

Mark   Certainly. For instance, several NP-complete problems are efficiently solvable on random instances. That is, there are very efficient algorithms that if the instance of the problem has been randomly chosen (among all instances of the same size) then, with high probability, the algorithm solves the problem or approximates the optimal solution with high accuracy. This phenomenon can be viewed as another aspect of the power of random choices. Here the random choices are embedded in the instance of the problem, while in the probabilistic algorithms they are part of the algorithm. As Karp[22] said, both aspects are important because although the probabilistic algorithms are more interesting, to date they are not able to cope with the explosion of combinations typical of NP-complete problems.

Laura   What you are saying does not exhaust either the phenomenon concerning difficult problems that admit "partially efficient" algorithms or the aspects relating to the random choices. In fact, there's a huge realm of algorithms whose behavior is often so complex as to make their mathematical analysis extremely difficult, and thus their performance is only evaluated through experimentation. These are usually called heuristic algorithms or simply heuristics, and they are developed to deal with difficult problems. Most of these heuristics use random choices. Just to name two among the most relevant: *simulated annealing* and *genetic algorithms*. For most heuristics it is even difficult to give just a rough idea of the types of instances on which they behave in an efficient manner.

Mark   That's right. In truth, the realm of heuristics is the "wildest" among those that belong to the world of algorithms, and it's also the one showing most clearly the weakness of current analytical techniques. We may be still very far from proving the truth or the falsity of the conjecture NP $\neq$ P and of many other conjectures of the theory of computational complexity. But even if we had all these demonstrations, it's not guaranteed that we would have the tools to understand which problems can be efficiently solved in practice and which not, with or without the help of random choices. In short, the power and limits of algorithms and random choices are very far from being understood, except perhaps for computability theory.[23] And since I came to make considerations on the ultimate frontiers of algorithms, the time has come for me to go away. I'm sorry, but I have to run.

Francis   Ah! Your words have charmed and numbed me. So, see you, bye!

Laura   Bye bye!

---

[22]Richard Manning Karp is one of the pioneers of the probabilistic analysis of algorithms and the theory of NP-completeness; he received the Turing Award in 1985.

[23]Computability theory, in essence, deals with the ultimate power of the algorithms. The main questions that it seeks to address are of the type: is there an algorithm (no matter how inefficient) that solves a given problem?

10.2 Bibliographic Notes

The conversation of the three friends has just touched the tip of the iceberg of probabilistic algorithms. Since they were introduced in the 1970s, their applications have proliferated: sorting algorithms, computational geometry, data mining, communication protocols, distributed computing, etc. The two books by Motwani and Raghavan [82] and Mitzenmacher and Upfal [80] deal in depth with probabilistic algorithms with regard to both the applications and the subtleties of their analysis.

The world of probabilistic algorithms is so vast and varied that even those two books together fail to capture it fully. The technique that has been called probabilistic counting is not covered in either of these books. An introduction to this interesting technique is contained in the paper [43].

Like probabilistic algorithms, the applications of hash functions are many and, as the conversation has shown, probabilistic algorithms and hash functions often go hand in hand. Virtually any book that introduces algorithms also treats the most common uses of hash functions. Crescenzi et al. [22] provides a lean and smooth introduction.

The three friends have discussed with animation the fascinating issues of primality testing and the factoring problem. One of the best books that addresses in detail primality tests (including that of Miller–Rabin), the most powerful factoring algorithms and their applications is [21]. The methods and algorithms used to generate pseudorandom numbers and the best statistical test beds to evaluate their quality are admirably presented and discussed in the second volume [69] of the monumental work by Knuth.

During the discussion, Kolmogorov complexity was invoked in relation to the impossibility of the existence of truly random generators. Actually, Kolmogorov complexity has ramifications that are far more extensive and has strong links with probabilistic methods. The previously mentioned [72] gives an introduction served with a rich collection of applications.

The intricate relationships between NP-complete problems, probabilistic algorithms, and random instances of hard problems are vividly recounted in the paper [65] by one of the fathers of the theory of NP-completeness. The even more intricate and delicate relationships among NP-completeness and, in general, computational complexity theory and the existence of algorithms that solve in the real world hard problems are open research issues that offer formidable difficulties and, maybe, for just this reason, have not yet been systematically studied. One of the very few papers addressing these issues and that gives an idea of this fascinating and unexplored land is [105].

References

1. AGCOM: Piano nazionale di assegnazione delle frequenze per la radiodiffusione televisiva. Autorità per le Garanzie nelle Comunicazioni (1998). http://www2.agcom.it/provv/pnf/target01.htm
2. AGCOM: Il libro bianco sulla televisione digitale terrestre. Autorità per le Garanzie nelle Comunicazioni (2000). http://www2.agcom.it/provv/libro_b_00/librobianco00.htm
3. Aho, A., Hopcroft, J., Ullman, J.: Data Structures and Algorithms. Addison-Wesley, Reading (1987)
4. Ahuja, R.K., Magnanti, T.L., Orlin, J.B.: Network Flows: Theory, Algorithms and Applications. Prentice Hall, Englewood Cliffs (1993)
5. Alpert, J., Hajaj, N.: We knew the web was big... Official Google Blog (2008). http://googleblog.blogspot.it/2008/07/we-knew-web-was-big.html
6. Aumann, R.J., Hart, S. (eds.): Handbook of Game Theory with Economic Applications. Elsevier, Amsterdam (2002)
7. Baeza-Yates, R.A., Ribeiro-Neto, B.: Modern Information Retrieval: The Concepts and Technology behind Search, 2nd edn. ACM, New York (2011)
8. Baeza-Yates, R.A., Ciaramita, M., Mika, P., Zaragoza, H.: Towards semantic search. In: Proceedings of the International Conference on Applications of Natural Language to Information Systems, NLDB 2008, London. Lecture Notes in Computer Science, vol. 5039, pp. 4–11. Springer, Berlin (2008)
9. Binmore, K.: Playing for Real. Oxford University Press, New York (2007)
10. Boyer, C.B., Merzbach, U.C.: A History of Mathematics, 3rd edn. Wiley, Hoboken (2011)
11. Brin, S., Page, L.: The anatomy of a large-scale hypertextual Web search engine. Comput. Netw. **30**(1–7), 107–117 (1998)
12. Broder, A.Z., Kumar, R., Maghoul, F., Raghavan, P., Rajagopalan, S., Stata, R., Tomkins, A., Wiener, J.L.: Graph structure in the Web. Comput. Netw. **33**(1–6), 309–320 (2000)
13. Buss, S.R.: On Gödel's theorems on lengths of proofs. II: lower bounds for recognizing k-symbol provability. In: Clote, P., Remmel, J. (eds.) Feasible Mathematics II, pp. 57–90. Birkhauser, Boston (1995)
14. Calin, G.A., Croce, C.: MicroRNA-cancer connection: the beginning of a new tale. Cancer Res. **66**, 7390–7394 (2006)
15. Cartocci, A.: La matematica degli Egizi. I papiri matematici del Medio Regno. Firenze University Press, Firenze (2007)
16. Chabert, J.L. (ed.): A History of Algorithms. From the Pebble to the Microchip. Springer, Berlin (1999)
17. Chakrabarti, S.: Mining the Web: Discovering Knowledge from Hypertext Data. Morgan Kaufmann, San Francisco (2003)

18. Cherkassky, B.V., Goldberg, A.V., Radzik, T.: Shortest paths algorithms: theory and experimental evaluation. Math. Program. **73**, 129–174 (1996)
19. Connes, A.: Visionari, poeti e precursori. In: Odifreddi, P. (ed.) Il club dei matematici solitari del prof. Odifreddi. Mondadori, Milano (2009)
20. Cormen, T.H., Leiserson, C.E., Rivest, R.L., Stein, C.: Introduction to Algorithms, 2nd edn. McGraw-Hill, Boston (2001)
21. Crandall, R., Pomerance, C.: Prime Numbers: A Computational Perspective. Springer, New York (2005)
22. Crescenzi, P., Gambosi, G., Grossi, R.: Strutture di Dati e Algoritmi. Pearson Education Italy, Milano (2006)
23. Davis, M.: The Universal Computer. The Road from Leibniz to Turing. W. W. Norton & Company, New York (2000)
24. Davis, M.: Engines of Logic: Mathematicians and the Origin of the Computer. W. W. Norton & Company, New York (2001)
25. Dawkins, R.: The Selfish Gene. Oxford University Press, Oxford (1979)
26. Demetrescu, C., Goldberg, A.V., Johnson, D.S.: The Shortest Path Problem: Ninth DIMACS Implementation Challenge. DIMACS Series. American Mathematical Society. http://dimacs.rutgers.edu/Workshops/Challenge9/ (2009). Accessed 15 Feb 2012
27. D'Erchia, A.M., Gissi, C., Pesole, G., Saccone, C., Arnason, U.: The guinea pig is not a rodent. Nature **381**, 597–600 (1996)
28. Devlin, K.: The Man of Numbers. Fibonacci's Arithmetic Revolution. Walker & Company, New York (2011)
29. D'Haeseleer, P.: What are DNA sequence motifs? Nature Biotechnol. **24**, 423–425 (2006)
30. Dijkstra, E.W.: A note on two problems in connexion with graphs. Numerische Mathematik **1**, 269–271 (1959)
31. Dijkstra, E.W.: The humble programmer. 1972 Turing Award Lecture, Commun. ACM **15**(10), 859–866 (1972)
32. Dijkstra, E.W.: This week's citation classic. Current Contents (CC), Institute for Scientific Information (ISI) (1983)
33. Dijkstra, E.W.: Appalling prose and the shortest path. In: Shasha, D., Lazere, C. (eds.) Out of Their Minds, The Lives and Discoveries of 15 Great Computer Scientists. Copernicus, New York (1995)
34. Divoky, J.J., Hung, M.S.: Performance of shortest path algorithms in network flow problems. Manag. Sci. **36**(6), 661–673 (1990)
35. Dowek, G.: Les metamorphoses du calcul, Une étonnante histoire de mathématiques. Le Pommier, Paris (2007)
36. EBU: Terrestrial digital television planning and implementation considerations. European Broadcasting Union, BPN 005, 2nd issue (1997)
37. Eco, U.: The Search for the Perfect Language. Blackwell, Oxford (1995)
38. Felsenfeld, G., Groudine, M.: Controlling the double helix. Nature **421**, 448–453 (2003)
39. Ferragina, P., Scaiella, U.: Fast and accurate annotation of short texts with Wikipedia pages. IEEE Softw. **29**(1), 70–75 (2012)
40. Ferragina, P., Giancarlo, R., Greco, V., Manzini, G., Valiente, G.: Compression-based classification of biological sequences and structures via the universal similarity metric: experimental assessment. BMC Bioinf. **8**, 252 (2007)
41. Ferro, A., Giugno, R., Pigola, G., Pulvirenti, A., Skripin, D., Bader, M., Shasha, D.: NetMatch: a Cytoscape plugin for searching biological networks. Bioinformatics **23**, 910–912 (2007)
42. Fetterly, D.: Adversarial information retrieval: the manipulation of Web content. ACM Comput. Rev. (2007). http://www.computingreviews.com/hottopic/hottopic_essay_06.cfm
43. Flajolet, P.: Counting by coin tossings. In: Proceedings of the 9th Asian Computing Science Conference, Chiang Mai, Thailand, pp. 1–12. Springer, Berlin (2004)
44. Ford, L.R. Jr., Fulkerson, D.R.: Flows in Networks. Princeton University Press, Princeton (1962)

45. Fredman, M.L., Tarjan, R.E.: Fibonacci heaps and their uses in improved network optimization algorithms. J. Assoc. Comput. Mach. **34**(3), 596–615 (1987)
46. Gallo, G., Pallottino, S.: Shortest path algorithms. Ann. Oper. Res. **13**, 3–79 (1988)
47. Garey, M.R., Johnson, D.S.: Computers and Intractability: A Guide to the Theory of NP-Completeness. W. H. Freeman, San Francisco (1979)
48. Giancarlo, R., Mantaci, S.: I contributi delle scienze matematiche ed informatiche al sequenziamento genomico su larga scala. Bollettino Della Unione Matematica Italiana – Serie A: La Matematica nella Società nella Cultura, 4-A (2001)
49. Giancarlo, R., Utro, F.: Speeding up the Consensus clustering methodology for microarray data analysis. Algorithms Mol. Biol. **6**(1), 1 (2011)
50. Goldberg, A.V., Harrelson, C.: Computing the shortest path: A* search meets graph theory. In: Proceedings of the Annual ACM-SIAM Symposium on Discrete Algorithms, Vancouver, Canada, pp. 156–165 (2005)
51. Golub, T.R., et al.: Molecular classification of cancer: class discovery and class prediction by gene expression. Science **289**, 531–537 (1998)
52. Graham, R.L., Hell, P.: On the history of the minimum spanning tree problem. Ann. Hist. Comput. **7**(1), 43–57 (1985)
53. Gusfield, D.: Algorithms on Strings, Trees, and Sequences: Computer Science and Computational Biology. Cambridge University Press, Cambridge (1997)
54. Gusfield, D.: Suffix trees (and relatives) come of age in bioinformatics. In: Proceedings of the IEEE Computer Society Conference on Bioinformatics, Stanford, USA. IEEE, Los Alamitos (2002)
55. Harel, D., Feldman, Y.: Algorithmics: The Spirit of Computing, 3rd edn. Addison-Wesley, Harlow (2004)
56. Hart, P.E., Nilsson, N., Raphael, B.: A formal basis for the heuristic determination of minimum cost paths. IEEE Trans. Syst. Sci. Cybern. **4**(2), 100–107 (1968)
57. Hawking, D.: Web search engines: part 1. IEEE Comput. **39**(6), 86–88 (2006)
58. Hawking, D.: Web search engines: part 2. IEEE Comput. **39**(8), 88–90 (2006)
59. Hinsley, F.H., Stripp, A. (eds.): Codebreakers: The Inside Story of Bletchley Park. Oxford University Press, New York (2001)
60. Hodges, A.: Alan Turing: The Enigma. Simon & Schuster, New York (1983)
61. Hood, L., Galas, D.: The digital code of DNA. Nature **421**, 444–448 (2003)
62. Horowitz, E., Sahni, S.: Fundamentals of Data Structures. Computer Science Press, Woodland Hills (1976)
63. Jones, N.C., Pevzner, P.: An Introduction to Bioinformatics Algorithms. MIT, Cambridge (2004)
64. Kahn, D.: The Codebreakers. Macmillan, New York (1967)
65. Karp, R.M.: Combinatorics, complexity and randomness. Commun. ACM **29**(2), 98–109 (1986)
66. Kaufman, C., Perlman, R., Speciner, M.: Network Security: Private Communication in a Public World. Prentice Hall, Upper Saddle River (2002)
67. Kleinberg, J., Tardos, É.: Algorithm Design. Addison-Wesley, Boston (2005)
68. Knuth, D.: The Art of Computer Programming. Volume 1: Fundamental Algorithms. Addison-Wesley Professional, Reading (1997)
69. Knuth, D.: The Art of Computer Programming. Volume 2: Seminumerical Algorithms. Addison-Wesley Professional, Reading (1998)
70. Lander, E.S.: The new genomics: global views of biology. Science **274**, 536–539 (1996)
71. Levitin, A.: Introduction to the Design and Analysis of Algorithms, 3rd edn. Addison-Wesley, Boston (2012)
72. Li, M., Vitányi, P.M.B.: An Introduction to Kolmogorov Complexity and Its Applications. Springer, New York (2008)
73. Li, M., Xin, C., Li, X., Ma, B., Vitányi, P.M.B.: The similarity metric. IEEE Trans. Inf. Theory **50**, 3250–3264 (2003)

74. López-Ortiz, A.: Algorithmic foundation of the internet. ACM SIGACT News **36**(2), 1–21 (2005)
75. Manning, C.D., Raghavan, P., Schutze, H.: Introduction to Information Retrieval. Cambridge University Press, New York (2008)
76. Mas-Colell, A., Whinston, M.D., Green, J.R.: Microeconomic Theory. Oxford University Press, New York (1995)
77. Matthews, W.H.: Mazes and Labyrinths. Longmans, London (1922)
78. Menezes, A., van Oorschot, P., Vanstone, S.: Handbook of Applied Cryptography. CRC, Boca Raton (1996)
79. Millennium problems. Clay Mathematics Institute. http://www.claymath.org (2000)
80. Mitzenmacher, M., Upfal, E.: Probability and Computing: Randomized Algorithms and Probabilistic Analysis. Cambridge University Press, Cambridge (2005)
81. Morelli, M., Tangheroni, M. (eds.): Leonardo Fibonacci. Il tempo, le opere, l'eredità scientifica. Pacini Editore, Pisa (1994)
82. Motwani, R., Raghavan, P.: Randomized Algorithms. Cambridge University Press, Cambridge (1995)
83. Nagel, E., Newman, J.: Gödel's Proof. NYU Press, New York (2008)
84. Nature Reviews: The double helix – 50 years. Nature **421** (2003)
85. Newson, M.W. (trans.): Mathematical problems. Bull. Am. Math. Soc. **8**, 437–479 (1902). (A reprint appears in Mathematical Developments Arising from Hilbert Problems, edited by Felix Brouder, American Mathematical Society, 1976)
86. Nisan, N., Roughgarden, T., Tardos, É., Vazirani, V. (eds.): Algorithmic Game Theory. Cambridge University Press, Cambridge (2007)
87. Orosius, P.: Historiarum Adversum Paganos Libri VII. Liber IV, 15. Thorunii (1857). Available online at The Library of Congress, call no. 7252181, http://archive.org/details/adversuspaganosh00oros
88. Osborne, M.J., Rubinstein, A.: A Course in Game Theory. MIT, Cambridge (1994)
89. Papadimitriou, C.H.: Computational Complexity. Addison-Wesley, Reading (1993)
90. Pavesi, G., Mereghetti, P., Mauri, G., Pesole, G.: Weeder Web: discovery of transcription factor binding sites in a set of sequences from co-regulated genes. Nucleic Acid Res. **32**, W199–W203 (2004)
91. Pizzi, C., Bortoluzzi, S., Bisognin, A., Coppe, A., Danieli, G.A.: Detecting seeded motifs in DNA sequences. Nucleic Acid Res. **33**(15), e135 (2004)
92. Pólya, G.: Mathematics and Plausible Reasoning. Volume 1: Induction and Analogy in Mathematics. Princeton University Press, Princeton (1990)
93. PTV: Planung Transport Verkehr AG (2009). http://www.ptvgroup.com
94. Rappaport, T.S.: Wireless Communications: Principles and Practice, 2nd edn. Prentice Hall, Upper Saddle River, New Jersey, USA (2002)
95. Rashed, R.: Al-Khwarizmi. The Beginnings of Algebra. Saqi, London (2009)
96. Reisch, G.: Margarita philosophica (1525) Anastatic reprint. Institut für Anglistik und Amerikanistik, Universität Salzburg (2002)
97. Sanders, P., Schultes, D.: Robust, almost constant time shortest-path queries in road networks. In: Proceedings of the 9th DIMACS Implementation Challenge Workshop: Shortest Paths. DIMACS Center, Piscataway (2006)
98. Scaiella, U., Ferragina, P., Marino, A., Ciaramita, M.: Topical clustering of search results. In: Proceedings of the Fifth International Conference on Web Search and Web Data Mining, Seattle, USA, pp. 223–232. ACM, New York (2012)
99. Shamir, R., Sharan, R.: Algorithmic approaches to clustering gene expression data. In: Current Topics in Computational Biology. MIT, Cambridge (2003)
100. Sharan, R., Ideker, T.: Modeling cellular machinery through biological network comparison. Nature Biotechnol. **24**, 427–433 (2006)
101. Silvestri, F.: Mining query logs: turning search usage data into knowledge. Found. Trends Inf. Retr. **4**(1–2), 1–174 (2010)
102. Simeone, B.: Nuggets in matching theory. AIRONews **XI**(2), 1–11 (2006)

103. Singh, S.: The Code Book: The Science of Secrecy from Ancient Egypt to Quantum Cryptography. Doubleday, New York (1999)
104. Stallings, W.: Cryptography and Network Security. Prentice Hall, Upper Saddle River (2007)
105. Stockmeyer, L.J., Meyer, A.R.: Cosmological lower bound on the circuit complexity of a small problem in logic. J. ACM **49**(6), 753–784 (2002)
106. Trakhtenbrot, B.A.: A survey of Russian approaches to perebor (brute-force searches) algorithms. IEEE Ann. Hist. Comput. **6**(4), 384–400 (1984)
107. van Lint, J.H.: Introduction to Coding Theory. Springer, New York (1998)
108. von Neumann, J., Morgenstern, O.: Theory of Games and Economic Behavior. Princeton University Press, Princeton (1944)
109. Wells, R.: Astronomy in Egypt. In: Walker, C. (ed.) Astronomy Before the Telescope. British Museum Press, London (1996)
110. Williams, J.W.J.: Algorithm 232 (heapsort). Commun. ACM **7**, 347–348 (1965)
111. Wirth, N.: Algorithms + Data Structures = Programs. Prentice Hall, Englewood Cliffs (1976)
112. Witten, I.H., Moffat, A., Bell, T.C.: Managing Gigabytes. Morgan Kaufmann, San Francisco (1999)
113. Witten, I.H., Gori, M., Numerico, T.: Web Dragons. Morgan Kaufmann, Amsterdam/Boston (2007)
114. Youschkevitch, A.: Les mathématiques arabes (VIII-XV siècles). Collection d'Histoire des Sciences, 2 – CNRS, Centre d'Histoire des Sciences et des Doctrines. VRIN, Paris (1976)
115. Zhang, S., Zhang, X.S., Chen, L.: Biomolecular network querying: a promising approach in systems biology. BMC Syst. Biol. **2**(1), 5 (2008)
116. Zobel, J., Moffat, A.: Inverted files for text search engines. ACM Comput. Surv. **38**(2), 1–56 (2006)